Real-Time
EARTHQUAKE
Tracking and Localisation

A Formulation for Elements in Earthquake Early Warning Systems (EEWS)

George R. Daglish
AND
Iurii P. Sizov

AuthorHouse™ UK
1663 Liberty Drive
Bloomington, IN 47403 USA
www.authorhouse.co.uk
Phone: 0800 047 8203 (Domestic TFN)
+44 1908 723714 (International)

Published by AuthorHouse 10/14/2019

ISBN: 978-1-5462-9682-9 (sc)
ISBN: 978-1-5462-9681-2 (e)

Library of Congress Control Number: 2018909612

Print information available on the last page.

Any people depicted in stock imagery provided by Getty Images are models,
and such images are being used for illustrative purposes only.
Certain stock imagery © Getty Images.

This book is printed on acid-free paper.

Because of the dynamic nature of the Internet, any web addresses or links contained in this book may have changed
since publication and may no longer be valid. The views expressed in this work are solely those of the author and do
not necessarily reflect the views of the publisher, and the publisher hereby disclaims any responsibility for them.

authorHOUSE®

Contents

Chapter 1

Introduction

The existence of powerful seismic events causing disaster, devastation, and severe dislocation is manifest.[1] This problem, which can deeply affect human society, is confronted from at least four vantage points:

- The epicentral/hypocentral location of earthquakes[2][3] – enabled by a global multiplicity of seismic station networks, international cooperation, and an understanding of the inner structure of the earth.
- The development of our understanding of the mechanisms that result in earthquakes and the direct observation of these.
- Direct observation at the confluence of tectonic plates by sea-going drilling platforms.
- Attempts to identify earthquake precursors.

It would appear vital for the future well-being of those who live in zones of high seismicity and potential high seismicity, for "Advanced seismological processing networks" (ASPNs) of some kind to be set up. Such networks could, in real, or near real time:

- Perform automatic epicentral/hypocentral localisations (with display and dissemimation of this information), by employing automatic detection of P and S arrivals, as well as surface wave onsets,
- Analyse and process incoming seismic signals to shed light on various models of earthquake mechanisms,
- Where the opportunity arises, correlate this signal analysis with knowledge gleaned from drilling activity—drilling though conjoined tectonic plates, as mentioned above.[4]

For long-range monitoring and possible prediction of minor and major earthquakes by such ASPN networks, several ranks, or levels, of data would appear to be required:

- The accurate extraction of the onset times for P, S, and various surface waves at seismic stations from seismic traces deconvolved from those output by receiving seismometers[5][6]
- The nature of those physical structures found within the scope of active zones and regions thought to contain mechanisms for earthquake generation
- The nature of those physical structures through which the various wave species emitted by the seismic events are to travel, together with their state.

---

[1]   RK McGuire, *Seismic Hazard and Risk Analysis*, EERI, 2004
[2]   WHK Lee, SW Stewart, *Principles and Applications of Micro-Earthquake Networks*, Academic Press 1981.
[3]   A Dziewonsky, *Seimology and the Structure of the Earth*, Elsevier, 2009
[4]   R Black, Reports on *Ocean-going Drilling Platforms: Ocean Monster Shows Hidden Depths*, BBC News, Japan, May, 2009.
[5]   J Havskov & G Alguacil, *Instrumentation in Earthquake Seismology*, Springer, 2004.
[6]   PS Kim, GL Oh, S Lu, A Yap, C Lock, *Detection and Classification of Acoustic and Seismic Events using a Semi-Markov Energy-dynamical Model,,* IMA Conferences on Mathematics in Defence, IMA 2009.

- Those parameters fundamental to models of earthquake mechanisms.[7]
- The state of such parameters
- And finally, the ongoing correlation of the above information with the time sequences of past earthquakes, within given zones of significant seismicity.[8] [9]

Prompted, particularly, by the great tsunami of 2004 and the 2008 Sichuan earthquake in China, we decided we should attempt to join these efforts by generating an experimental seismic software system, which should contain facilities, or application areas, that we could work up for inclusion as components in an earthquake early warning (EEW) software structure.

The body of *Real-Time Earthquake Tracking and Localisation* represents the state this experimental system has so far reached.

We first drew up a high-level project specification to give some direction and backbone to our research.

Specification

Those considerations set out initially call for research and development aimed at providing the consolidated theoretical background that should ultimately enable the implementation of such large intelligent seismic array systems with wide aperture.

The range of earthquakes that such systems would have to perceive, analyse and, locate should range from micro- and ultra-micro earthquakes $(M_L < 3)$ to small and moderate earthquakes $(3 \leq M_L < 7)$ and to major earthquakes $(M_L \geq 7)$. Here M_L represents local magnitude in the Richter/Gutenberg sense. Further measures for magnitude include:

1. Surface wave magnitude (Gutenburg): $M_S = \log A - \log A_0 \left(\Delta^\circ \right)$

2. Body wave magnitude (Gutenburg): $m_b = \log \left(\frac{A}{T} \right) - f \left(\Delta, h \right)$

3. Generalised magnitude (Bath): $M = \log \left(\frac{A}{T} \right) + C_1 \log \left(\Delta \right) + C_2$

4. Duration magnitude (Suteau and Whitcomb): $M_D = a_1 + a_2 \log \tau + a_3 \Delta + a_4 h$

The dynamic range of seismic recording can be such that different families of stations should be provided to sense earthquakes at the aforementioned and correspondingly different bands of magnitude.

As stated at the outset, the need for such seismic arrays is manifest. They are required to monitor seismicity in regions where serious earthquakes occur on a regular basis. Further, they should provide a monitoring capability on a global scale both for natural and man-made seismic events.

---

[7]   J. Koyama, *The Complex Faulting Process of Earthquakes*, Kluwer Academic Press, 1997.

[8]   R Saladin & B Vermeersen, *Global Dynamics of the Earth*, Springer, 2004.

[9]   P Keary, *Global Tectonics*, Wiley-Blackwell, 2009.

Such a network should be available to study patterns of seismic activity, foreshock activity, P-wave velocity, velocity anomalies, and more. Statistics can then be gathered and built up to evaluate, for instance, any predictive/descriptive formulae derived from the Ishimoto–Iida relation. For example,

$$\log N = a - bM; \quad (b \approx Unity, \text{ by observation})$$

The original Ishimoto–Iida relation was discovered as

$$NA^m = k \quad (\text{const})$$

Here, N is the frequency of occurrence, and A is the maximum trace amplitude for earthquakes at approximately similar focal distances. m is found empirically to be 1.74. The number (N) of earthquakes is seen to increase tenfold with unit decrease in magnitude. Therefore, the greater number of smaller magnitude events one can collect, the greater knowledge of the occurring seismicity patterns one may obtain.

Precursor concepts include the LASA (Large Aperture Seismic Array) in Montana, United States of America; the NORSAR in Norway, Scandinavia; the World-Wide Standardized Seismic Network (WWSSN), along with the Global Seismic Network (GSN), established by Incorporated Research Institutes for Seismology (IRIS) and the USGS Central Californian Microearthquake Network.

The projected research program, as envisaged, could be split into two phases:

1. Exploration of the basic concepts and their validation in small- and large-scale simulation
2. Validation of the basic concepts by constructing simulated prototype seismological systems that emulate real-time working

Phase 1: Exploration of the Basic Concepts and their Validation in Small- and Large-Scale Simulation

This initial body of research is seen as having at least the following five components:

1. Development of the simple spherical shell or lamina and then the solid sphere, with regard to the solution of these simplified equation systems, in particular

$$\underline{x} \cdot \underline{a_i} = R^2 \cos\left(\frac{\overline{c}_0(t_i - \delta t)}{R}\right); \quad i = 0, (n-1) \qquad (1741,1)$$

$$S_n = \sum_{i=0}^{n-1}\left(\sqrt{u_i} - \overline{c}_0(t_i - \delta t)\right)^2 \qquad (1,2)$$

where

$$u_i = \left|\underline{x} - \underline{a_i}\right|$$

and

$\underline{x}$ is the location of the Seismic Emission in cartesians.

$\underline{a_i}$ is the location of the ith sensor in cartesians.

R is the radius of a spherical shell.

t_i is the ith arrival time relative to the lead sensor time.

δt is the time-to-origin, i.e. the arrival time at the lead sensor.

$\overline{c_0}$ is, in some sense, an averaged wave propagation velocity.

Then the solution vector for both (1.1) and (1.2) is:

$$\underline{s} = \begin{bmatrix} \underline{x} \\ \delta t \\ \overline{c_0} \end{bmatrix}$$

The Cartesians referred to here are coordinates within a space frame whose origin coincides with that of the centre of the object sphere. Thus, in (1.1),

$$\|\underline{x}\| \approx R; \quad \|\underline{a_i}\| \approx R.$$

This fact allows the following derivation:

An arc length is given by an expression such as $R\theta$ where R is the radius of the sphere or circle in question. Therefore, considering the sphere on whose surface emissions are being transmitted and removing the coordinate system origin to the centre of this sphere, we get

$$R\cos^{-1}\left(\frac{xa_i + yb_i + zc_i}{\|\underline{x}\| \cdot \|\underline{a_i}\|}\right) = \overline{c_0}(t_i - \Delta t); \quad \iota \in \left[0, (n-1)\right],$$

But :

$$\|\underline{x}\| \cdot \|\underline{a_i}\| = R^2.$$

So :

$$R\cos^{-1}\left(\frac{xa_i + yb_i + zc_i}{R^2}\right) = \overline{c_0}(t_i - \Delta t).$$

Transposing functions gives

$$xa_i + yb_i + zc_i = R^2 \cos\left(\frac{\overline{c_0}(t_i - \Delta t)}{R}\right); \quad i \in \left[0, (n-1)\right],$$

as above. This system is then solved for $\left(x, y, z, \Delta t, \overline{c_0}\right)$.

To fix $\underline{x}$ use this system, $n \geq 5$. It is notable that the above system is in the form of a set of planes the lengths of whose normals to the origin are time dependent and oscillate as cosine functions. We can write therefore for one such plane

$$\underline{a_i} \cdot \underline{x} = f(t_i)$$

$$f(t_i) = R\cos\left(\frac{\overline{c_0}}{R}(t_i - \Delta t)\right)$$

$$f(t_i) \in [-R, R]$$

Here, $\underline{a}$ is now a vector of direction cosines for the ith normal. And $\underline{x}$ is the Cartesian point of emission in the space frame within which the sphere is embedded and whose origin coincides with the centre of the sphere.

These direction cosines will be given by

$$\underline{a_i} = \frac{(a,b,c)_i}{R}$$

A wave front corresponding to a specific velocity, $\overline{c_0}$ (in some sense a mean value for an isotropic medium) is defined by the intersection of the oscillating plane and a sphere of radius R.

It is also of note that the period of oscillation for the entire set of planes is given by

$$T = \frac{2\pi R}{\overline{c_0}}$$

The restitution of $\underline{x}$ can be considered to represent the point where all n planes mutually intersect and where, indeed, the point of intersection lies on the spherical surface of radius R.

The first of these equation systems (1.1) represents transmission over the surface of a spherical lamina. The second system represents a simple expression for transmission through the body of a solid and, ultimately, an arbitrarily layered sphere (Geiger's method).

The Geiger concepts may be invoked so that either the earthquake focus (hypocentre) is found or estimated using a simple radial velocity structure or a simultaneous inversion problem can be solved, establishing the hypocentre parameters, using a complex three-dimensional velocity structure.

2. Investigation of the properties of (micro) seismic events in terms of the wave trains emitted as acoustic emissions – Investigation will include the following wave types:

P and S waves
Rayleigh waves
Love waves
Lamb waves

The use of initial system, above, allows for the analysis of three questions. First, can surface wave types (Rayleigh and Love waves), which are prevalent in certain earthquakes, be transmitted in such a way that this system may be deployed to determine epicentres (or mesocentres) from very long range to close up? Second, is there a mechanism that converts P and S waves to surface waves of this type when they impinge on the surface of the crust? Third, what is the occurrence and range of the approximation of the ducting of P and S waves by the lithosphere and is this a truly viable means of finding the epicentre using this system ?

These questions can be summed up as follows: What are the interrelationships between this group of waveforms and which forms can be processed by this system to establish seismic event parameters? Also how does directed body wave energy manifest itself to a surface or interface?

3. Projected properties of the acoustic emissions (AE) on and within spherical lamina in terms of

   a) total internal reflection and
   b) interference patterns

Both will be governed by the refractive properties, the radius of curvature, and the laminar thickness of the body. With regard to the velocity calculations, then, with ducting, the slowest path is that which progresses with reflections made at the critical angle.

The attenuation and bandpass effects of the "earth system" (surface, stratification, interfaces, and so on) on the body and surface waves generated by seismic events should also be investigated with a view to establishing an optimal spatial distribution for any projected layout of a system.

4. A study of the means whereby automatic (in other words, hands off) arrival time detection and cut-off time definition is to be achieved, given the characteristics of the wave patterns so formed – There are many suggestions to be taken from successful monitoring systems for this component.

One basic observation is that, given an onset of any of these wave types, in particular the potential arrival of P energy, then we might say

$$\alpha_P(\tau_\alpha) = \frac{1}{\tau_\alpha} \int_{t-\tau_\alpha}^{t} |A(t)| \, dt$$

$$\beta_P(\tau_\beta) = \frac{1}{\tau_\beta} \int_{t-\tau_\beta}^{t} |B(t)| \, dt$$

$$\gamma = \frac{\alpha_P}{\beta_P},$$

This is onset detection by energy considerations, and the following definitions apply:

   $A(t)$ the train of displacements due to the potential P-energy

   τ_α the time period over which the integration of the displacements takes place

   $B(t)$ also the train of displacements by the potential P-energy

   τ_β the time period over which this latter integration takes place.

We note that $A(t) \supset B(t)$ and that $\tau_\alpha < \tau_\beta$.

If, then, the ratio "γ" exceeds a prescribed threshold $(\geq Unity)$, an onset is deemed to have occurred.

However, we must also take into account new work instigated and developed by Pang Sze Kim et al. (13), which detects events in energy profiles using particle filters.

A further question is, Could we elevate the concept of "motion intensity" to the status of a pure onset indicator?

5. A study of the optimal manner in which to process all arrival times to (a) produce locations of hypocentres, epicentres, and mesocentres, together with their parameters, and (b) establish velocity structures for the region.

This will also involve the processing of event swarms. These aspects will include:

a.   Arrival time buffering
b.   Combinatorial scans in real time using cross-correlation, semblance functions and the like to ease the computational load
c.   Estimation of error and further discriminating processes and including several possible calculation types such as those given above.

Further analysis, as distinct from the location calculation, should take place as background computation. Such analysis would include:

a.   Determination of the Travel Paths with radial velocity fields, as well as 3D velocity fields
b.   Spatial derivatives
c.   Take-off angles
d.   Fault plane solution

With regard to the seismic moment (M_0), then it should be verified that

$$M_0 = \left(\frac{\Omega_0}{\psi_{\theta\phi}}\right) 4\pi\rho R v^3$$

It should also be seen how M_0 can be obtained from surface waves and coda waves.

Here,

Ω_0 is the long period limit of the displacement spectrum of the P- or S-energy

$\psi_{\theta\phi}$ represents the body wave radiation pattern

R is a function accounting for the spreading of body waves

v here, is the body wave velocity.

The seismic moment (1966) has been introduced as

$$M_0 = \mu \int_A D(A)\, dA,$$

where

μ is the shear modulus of the medium

A is the area of the slipped surface or the "source area"

$D(A)$ is the slip.

In this, we can see that the seismic moment is a direct measure of the strength of an earthquake caused by fault slip. Therefore, the computational procedures for fixing the fault plane from P-wave first motion data should be investigated.

Further means of calculating the earthquake parameters may be gained from spectral analysis of the incoming body waves.

Phase 2: Validation of the Basic Concepts by Constructing Prototype Seismological Systems in the Lab

Code resulting from testing and simulation in Phase 1 would form the kernel for the design of the software systems to be produced as:

1. The design of the structure and the actual software for the processing of the telemetered data from the seismic stations must provide information on any seismic event perceived by this system.
2. Each seismic station must act as a remote data capture system, which uses onset recognition and cut-off to transmit the requisite seismograph segments.
3. The software will have the characteristics of a multi-layered real-time acquisition and processing system.
4. With regard to performance, the system should be designed to cope with event swarms, both by buffering of input and by swift processing techniques.
5. The testing of the finished prototype system will use concrete examples taken from actually occurring earthquakes.

Chapter 2

Guidelines for Prototype Seismic System (EEW)

This chapter acts as an introduction to the prototype epicentre and hypocentre location system for EEW.

There would appear to be two main approaches to gaining the localisation parameters of an earthquake—epicentre $(\underline{\varepsilon})$ and hypocentre H_d:

I. Gain an epicentre and scan for the hypocentre.
II. Concurrently scan for epicentre and hypocentre using tabular sets of P wave (or S wave) first arrival travel-times, based on any chosen earth velocity model (radially spherical).

In approach I, the epicentre can be found by:

1. P and S wave time difference on an approximately "flat" earth
2. P and S wave (or Love and Rayleigh wave) time difference on a truly spherical earth
3. Any pair of first arrivals from the set of P, S, Love, or Rayleigh waves on a truly spherical earth
4. The direct calculation for a "surface source" with primarily Love and Rayleigh waves (and possibly P or S waves) using

$$\underline{x} \cdot \underline{a}_i = R^2 \cos\left(\frac{\bar{c}_0\left(t_i - \delta t\right)}{R}\right); \quad i = 0, (n-1)$$

as given above.

In approach I, the hypocentre scan can be either (a) an "explicit" scan using a particular choice of point-to-point (P2P) ray tracer, combined with a given earth velocity model or (b) an "interpolative tabular scan." where the tables of first arrival P-wave (or S-wave) energy have been produced by, again, a choice of ray tracer and earth velocity model.

In approach II, the concurrent scan for epicentre and hypocentre is self-contained, but the hypocentre result can be refined by any method.

In implementing techniques for localisation, we have worked through all the above types and headed for the major type II as being the most likely candidate for a front-end software component in an EEW system

A good deal of the material presented in this volume is not new but represents a fundamental basis that is needed to inform the development of the software.

In the description of the intended software structure, there are the following entities (colour-coded as below):

1. **Data/Information**
   - Text files
   - Console or screen input
   - Graphical output
     (a) Depiction of seismograms
     (b) Depiction of energy onset within individual seismograms

2. **Programs/System Nodes**
   These entities, when activated, cause the general flow of data within the system. They are supported by the following groups of functions:
   - **EarthQuakeLocationHeader.h**
   - **EikonalScanningTools.h**
   - **EpicentreLocationHessian.h**
   - **FFTandCorrelationD.h**
   - **FFTandCorrelationF.h**
   - **HypocentreConsolidationCC.h**
   - **HypocentreConsolidationLL.h**
   - **HypocentreScanningTimeBased.h**
   - **HypocentreScanningTimeBasedD.h**
   - **HypocentreScanningTimeBasedF.h**
   - **HypocentreScanningTransBased.h**
   - **HypocentreScanningTools.h**
   - **IntegrationTools.h**
   - **QuakeLocationGaussNewton.h**
   - **QuakeLocationHeader00.h**

3. **External Systems**
   - WILBUR II or III (IRIS)
   - EXCEL 2005 (Microsoft)

   Wilber 2 (and its beta version, Wilber 3), from the Incorporated Research Institutions for Seismology (IRIS), is a means whereby users from within IRIS can extract seismological data for their own research purposes.

   Microsoft's Excel 2005 is here used, under the Windows XP operating system, to generate graphical output. This will contain cues that can ultimately form input to the system from the screen/ console and so may be crucial to some attempts by the system to locate epicentral coordinates.

   Entity and pathway definitions are given both in Annex A and here. Each pathway described here consists of a set of program nodes, described in Annex A.

Canonical Pathway (I)

It can be seen by looking at the information above that the basic pathway through this system (you might say the critical path) follows the following route:

1. RDTST 04, 05 or 06, 07, and then 02 (input seismograms to system)
2. FSTST 02 (perform FFT on each input seismogram)
3. IGTST 05, 07 (integrate in the frequency domain, using FFT)
4. ISTST 00 (gather individual integrated seismograms)
5. PWTST 05, 07, 09 or 10 (detect energy onsets within seismograms)
6. PWTST 06 (refine detection of energy onsets)
7. IPTST 00 or 01 (extract and input energy onsets)
8. HCTST 00 (preset station parameters for ETTST 01 or 08)
9. DLTST 01, DPTST 01 or DQTST 01 (epicentre from energy onset arrival differences)
10. AUTST 01, 02, 03 or 04, 06 (define arrivals to pass to location routines)
11. DynamicSphericalDeterminationAvto 03, 04 (scanning epicentral location routine)
12. StaticSphericalDeterminationAccelerate (epicentral location routine) or StaticSphericalDeterminationBypass (data selection and move)
13. EarthQuakeLocation07, 08 or 09 (Final hypocentral location scanning routines directly using P2P ray tracers)
14. ETTST 01 or 08 (final hypocentral location using tabular scanning routines)
15. UETST 00 or UETST 01 (Further examination of the accuracy of the epicentre/hypocentre location)

Canonical Pathway (II)

This second pathway is extremely brief. It is based on routines that need only the P-wave first arrivals to act. It thus obviates any picking of S-wave, L-wave, and R-wave species. It would appear to be germane to earthquake early warning (EEW) software regimes. It follows the following course, which is shown diagrammatically in chapter 5:

1. RDTST 04, 05 or 06, 07 and then 02 (input seismograms to system)
2. RQTST 01 or 02 (01 transforms the set of station coordinates for use in scan)
3. HCTST 01 or 02 (presets station parameters and timings for subsequent use in location)
4. SQTST 01, 02a or 03 (each routine locates epicentre and hypocentre simultaneously)

The output of this pathway can be acted upon for refinement by the scanning routines ETTST 01 or 08 or any of the latter end routines on the first pathway that will provide such refinement.

Canonical Pathway (III)

This third pathway performs a reduced simulation for individual earthquakes. The simulation models the tracking of earthquakes by the concurrent localisation of their epicentres and hypocentres, using the first arrivals for the P waves (SQTST 06) as they spread out and incrementally impinge on elements of a given set of sensors. For each run of this pathway, the data is fed to the algorithm station by station. In this case, no attempt is made to actively model and emulate real-time execution. However, attention is paid to the time lapses between the points at which P waves impinge on ("illuminate") elements of the set of stations. Briefly we have:

1. RDTST 07 and then 02 (extracts information from a given set of SACA seismograms and their headers)
2. RQTST 02 (creates a list of absolute sensor coordinates and first arrival P-wave timings
3. HCTST 02 (passes on a cumulative table of station coordinates and their corresponding P-wave arrival times)
4. SQTST 06 (takes subsets of chronologically ordered data from the chosen data structure, generated at 3)
5. ETTST 10 (is capable of making a final refinement, or confirmation, of the hypocentre value and set of take-off angles, given the value for the epicentral coordinate vector provided at 4)

The full structure for all varieties of this pathway is given in chapter 5.

Canonical Pathway (IV)

This is again a reduced simulation, and a fuller description is given in Annex E. Pathway IV consists of any one of the set of routines

ETTST 14.cpp to ETTST 20.cpp.

This would now include especially

ETTST 16.cpp & ETTST 16D.cpp to ETTST 16EZ.cpp

These routines contain the structure and functions of points 4 to 5 in Canonical Pathway III. They use input for particular earthquakes created by routines at points 1 to 3 in this pathway.

ETTST 16.cpp, ETTST 16D.cpp ETTST 16EX, ETTST 16EY.cpp & ETTST 16EZ.cpp contain the latest distillation of the Canonical Pathway (IV).

We should also note that the system as a whole is supported by the above-mentioned groups of subroutines or functions.

Apart from point-to-point (P2P) ray tracers, integration routines, matrix inversion, and non-linear optimisation, there are various data retrieval and coordinate manipulation procedures, all of which are vital for the functioning of the system as a whole and which are to be the subject of a further article of documentation.

To each of the above pathways we can also add a prelude:

1. TimeBased01 – Generate scanning tables, which consist of
   • Travel times
   • Take-off angles

- Calibration data
- Ray path lengths
- Accuracy level data
- Error log.

Each file, except the last, contains data in the form of a matrix, which is formed from a discretised two-space field whose dimensions represent relative co-latitude and depth in terms of columns and rows. The last item is a list of depths and target co-latitudes for which the point-to-point ray tracer did not achieve the required level of convergence.

2. MXTST 00, MQTST 00, and MSTST 00 – Nodes ferry the tables generated at 1 to their storage locations within the system and, from that point, to the locations from which they are actively input to the scanning routines.

3. ETTST 04, 05, 06, 07 – Adjust and compensate for error in table generation by interpolation and using a selection of L/S filters to smooth and to infill where it becomes necessary.

Overall system diagrams corresponding to general canonical pathways are given at Annex B.

Chapter 3

Mathematical Discussion

This description should be read in conjunction with Annex A and the system diagrams given at the end of that annex.

At the heart of this prototype location system lie several families of algorithms:

1. Deconvolution of raw seismic input, using individual seismometers impulse responses
2. Integration of accelerogram input or velocigram input using FFT techniques implemented in FTTST 02 and in IGTST 05
3. Energy onset detection by PWTST 05 and PWTST 06 and so on.
4. Estimation of epicentral coordinates from time differences at individual stations by DLTST 00 and by other means, as in DPTST 00 and DQTST 00.
5. Direct estimation of the epicentral coordinates as the vector

$$\underline{x} = \begin{bmatrix} x \\ y \\ z \\ \Delta t \\ \overline{c}_0 \end{bmatrix}.$$

being a solution of the system

$$xa_i + yb_i + zc_i = R^2 \cos\left(\frac{\overline{c}_0\left(t_i - \Delta t\right)}{R}\right); \quad i \in \left[0, (n-1)\right],$$

to be discussed below.

6. Determination whether or not the epicentres so found are "oceanic"—in other words, coastal (inland) or maritime
7. A scan for hypocentre, involving a family of ray trace algorithms
8. Drawing of inferences concerning the given fault plane mechanism
9. Modelling of the tsunamis that may thus ensue, to give first arrival times
10. Deployment of "intelligenced" software to look for event precursor patterns

The estimation at 5 above is performed by the routines

DynamicSphericalDeterminationAvto

and

StaticSphericalDeterminationAccelerated.

In $\underline{x} = (x, y, z)$ is the Cartesian position of the epicentre in a space frame, rotating with the reference sphere, whose radius is R. The origin of this space frame coincides with the centre of the reference sphere. The set $\{\underline{a}_i\}$ represents the coordinates of the seismographic stations in the same frame of reference, while the set $\{t_i\}$ represents the set of onset timings taken relative to the lead station, such that $t_0 = Zero.$, Δt, and $\overline{c}_0$ are the time to origin and an estimate of wave front velocity, respectively.

Impulse Response: Deconvolution

The exercise here is to perform the following algorithm:

1. Generate an impulse response as a time series by first solving the equation

$$\frac{d^2 z}{dt^2} + K_1 \frac{dz}{dt} + K_0 z = \delta(0)$$

This becomes

$$z(t) = \frac{1}{(a-b)} \cdot \left(e^{at} - e^{bt}\right)$$

where

$$(a-b) = \sqrt{K_1^2 - 4K_0}$$
$$a = \frac{-K_1 + \sqrt{K_1^2 - 4K_0}}{2}$$
$$b = \frac{-K_1 - \sqrt{K_1^2 - 4K_0}}{2}$$

2. Use the function $z(t)$ to provide a Laurent's series:

$$Z(h_i) = \sum_{i=0}^{\infty} z(t_i) z^{-i}$$

3. Perform an inversion on this Laurent's series to give

$$\frac{1}{Z(h_i)}$$

4. Convolve an observed waveform with the inverted Laurent's series from 3, thus reconstituting an input, or unobserved incoming waveform.

Although it is germane to the deconvolution of observed waveforms to give a version of the incoming wave, this routine has not yet been factored into the present scheme. It will act as an intermediary between the RDTST dd set and its following node, FTTST 02.

George R. Daglish and Iurii P. Sizov

Integration of Seismogram Input

The principal aim behind this subsystem is to accurately determine the inner structures of the object seismograms by "frequency band splitting" and examining the resultant integrated waveforms within each band. This may result in some specific bands showing more clearly than others where the Love/ Rayleigh energy onsets are occurring. In the current tests, for instance, using information from the Chilean earthquake (May 2010), the bands (hertz)—(0.01, 0.1); (0.1, 1.0); (38, 40)—when integrated to the second level (motion) give good and relatively clear indications of such energy onsets.

The calculation regime employed to integrate from accelerogram to velocity, and also to motion and displacement, is performed in the frequency domain by IGTST 05 (using coefficients generated by FTTST 02) according to the relationship, depicted by the scheme

$$\int^n \int \int \int \ldots \int f(t).dt \leftrightarrow \left(\frac{1}{j\omega}\right)^n \cdot F(\omega)$$

or, in more general terms,

$$\int^n \int \int \int \ldots \int f(\tau) \cdot d\tau \leftrightarrow \frac{(-j)^n}{\omega^n} \cdot F(\omega)$$

In our case, we have (although there is a comprehensive series for continued integration)

$$\int f(t) \cdot dt \leftrightarrow \left(\frac{1}{j\omega}\right) \cdot F(\omega)$$

and

$$\int\int f(t) \cdot dt \leftrightarrow \left(\frac{1}{j\omega}\right)^2 \cdot F(\omega)$$

In this, $f(t)$ represents the input accelerogram at a given sample rate, and $F(\omega)$ represents the initial Fourier transform of this function. It may be that the accelerogram has been deconvolved from the particular seismometer's impulse response function.

We can see that, with respect to the discrete Fourier transform (DFT) or the fast version of this, the FFT (fast Fourier transform)—due latterly to the Cooley–Tukey algorithm—we have the following operational sequence:

1. The accelerogram is submitted as an entirely real series with

$$P(i) = f(t_i) \wedge Q(i) = Zero$$

or

$$(P_t + jQ_t)_i = f(t) + j\underline{0}$$

We say

$$Pc = \sum_k P_k \cos\left(\frac{2\pi}{N} \cdot i \cdot k\right)$$

$$Ps = \sum_k P_k \sin\left(\frac{2\pi}{N} \cdot i \cdot k\right)$$

$$Qc = \sum_k Q_k \cos\left(\frac{2\pi}{N} \cdot i \cdot k\right)$$

$$Qs = \sum_k Q_k \sin\left(\frac{2\pi}{N} \cdot i \cdot k\right)$$

This is the output from the Fourier transform, where the input is multiplied by $(c - js)$. In addition,

$$Pc = \sum_i P_i \cos\left(\frac{2\pi}{N} \cdot i \cdot k\right)$$

$$Ps = \sum_i P_i \sin\left(\frac{2\pi}{N} \cdot i \cdot k\right)$$

$$Qc = \sum_i Q_i \cos\left(\frac{2\pi}{N} \cdot i \cdot k\right)$$

$$Qs = \sum_i Q_i \sin\left(\frac{2\pi}{N} \cdot i \cdot k\right)$$

is the output from the Fourier transform, where the input is multiplied by $(c + js)$.

In this context, $(c \pm js)$ is short for $e^{\pm j\alpha_i}$.

The output is held as a complex series in frequency as

$$\underline{P}(\omega) + j\underline{Q}(\omega)$$

2. If we wish to re-synthesise the initial input from its transform without invoking the very slow process of the explicit IDFT (inverse discrete Fourier transform), we can resubmit the output to the FFT algorithm as a complex conjugate. This will generate (given a scaling factor) the original real-time series of the accelerogram. We have provided

$$P_i + jQ_i$$

Taking the complex conjugate of the coefficients output by the FFT and re-inputting to the FFT, as an inverse transform, we get

$$(P - jQ)(c - js)$$
$$\Downarrow$$
$$(Pc - Qs) - j(Ps + Qc)$$

This latter represents, in its real part, the initial input function, $f(t_i)$, while the imaginary part remains at $\underline{0}$. Now, taking this latter output as, again, a complex conjugate and again re-inputting to the FFT as a complex conjugate and again re-inputting to the FFT as an inverse process, we get

$$\{(Pc-Qs)+j(Ps+Qc)\}(c-js)$$

$$\Downarrow$$

$$P+jQ$$

which is the set of coefficients initially generated from the input of the series $f(t_i)$.

Thus we see that this series, which can be formed from the FFT as an inverse process from its frequency space coefficients, when used as data to a forward FFT process, will generate its originating frequency space coefficients. This applies to levels 1, 2, and 3 of integration, as it does for level 0, as is shown above. We now take each of this set of integration levels in turn.

This validates the assumption that we can integrate within the frequency space and, by using the forward FFT as an inverse transform, resurrect the required integral in the time domain.

The rule, simply stated, is:

Always re-input (to the FFT routine, which is to act as an inverse transform) the complex conjugate of the level 0 frequency output, frequency-integrated to the required level (0, 1, 2, or 3, in this case) to regain the required function in the time domain, as integrated to this required level.

3. Following a similar rationale, we integrate the initial complex output of the FFT as

$$\left(\frac{1}{j\omega}\right)(P+jQ)$$

$$\Downarrow$$

$$(Q-jP)$$

This is re-input as conjugate to the FFT

$$(Q+jP)(c-js)$$

$$\Downarrow$$

$$(Qc+Ps)+j(Pc-Qs)$$

which is the integral of the $f(t_i)$ sequence,

$$v(t_i)=\int_0^{t_i} f(\tau)\cdot d\tau$$

Re-inputting the conjugate of this series to the FFT, we recover the initial complex coefficients

$$\{(Qc+Ps)-j(Pc-Qs)\}(c-js)$$
$$\Downarrow$$
$$Q-jP$$

4. Integrating in a similar manner, generate the motion as

$$s(t_i)=\iint_{t_i} f(\tau)\cdot d\tau$$

We get

$$\left(\frac{1}{j\omega}\right)^2 (P+jQ)=\frac{1}{\omega^2}(-P-jQ)$$

Inputting to the FFT, we get the series for $s(t_i)$ as

$$(-P+jQ)(c-js)$$
$$\Downarrow$$
$$-(Pc-Qs)+j(Ps+Qc)$$

We now again input this as a complex conjugate to the FFT procedure and retrieve the initial set of coefficients,

$$\{-(Pc-Qs)-j(Ps+Qc)\}(c-js)$$
$$\Downarrow$$
$$-P-jQ$$

5. Again integrating to gain the function

$$a(t_i)=\iiint_{t_i} f(\tau)\cdot d\tau$$

we have

$$\left(\frac{1}{j\omega}\right)^3 (P+jQ)=\frac{1}{\omega^3}(-Q+jP)$$

Input the conjugate to the FFT

$$(-Q - jP)(c - js)$$

$$\Downarrow$$

$$-(Qc + Ps) + j(Qs - Pc)$$

This forms the required series, $a(t_i)$, above. The conjugate of this series, when input to the FFT will bring back the original frequency domain coefficients as

$$\{-(Qc + Ps) - j(Qs - Pc)\}(c - js)$$

$$\Downarrow$$

$$(-Q + jP)$$

Employing the FFT, and avoiding a literal IDFT, leads to very rapid processing. On a 3.2 gigahertz machine, we can achieve the single or double integration for sequences of 2^{17} samples with equal times of within $d.ddddd_{10} - dd$ seconds.

Further, using the initially generated $z_i = P_i + jQ_i$, we can add a bandpass selection of frequencies to input to the FFT integration processes. This enables us to clearly see how the contribution of specific frequency bands forming the wave structures emanating from the seismic event in question is made. Some examples of this are shown in Annex D.

This process could be referred to as frequency band splitting, where discrete bands are chosen with specific widths for each band.

However, a more ideal procedure would be represented if the width of each instantaneous band were selected dynamically while the basal frequency of the "band window" was made to slide though some subset of the frequencies present in the accelerogram. For each band window and its basal frequency, the required integration should take place. Such a procedure would allow a complete review of those resultant waveforms most suited to demonstrate the arrivals of the set of P- S-, L-, and R-wave species for the ultimate purposes of their event location.

Energy Onset Detection

The subject of detecting energy onset has two basic approaches:

1. Using the FFT on the seismogram to determine the power spectra for the two time segments that together make up the detection scheme
2. Integrating the absolute values of the seismogram sample data to obtain an indication of the mean power values for the two time segments forming the detection mechanism

The first method can be slower than the trapezoidal integration that may be used in the second method. (The latter, in fact, proves to be much more rapid.) The essence of both methods is given by this general consideration—there are many suggestions to be taken from systems that successfully monitor energy onset.

Take this one basic observation. Given an onset of any of the wave types (in particular the potential arrival of P energy), then we might say

$$\alpha_P\left(\tau_\alpha\right) = \frac{1}{\tau_\alpha}\int_{t-\tau_\alpha}^{t}\left|A(t)\right|dt$$

$$\beta_P\left(\tau_\beta\right) = \frac{1}{\tau_\beta}\int_{t-\tau_\beta}^{t}\left|B(t)\right|dt$$

$$\gamma = \frac{\alpha_P}{\beta_P},$$

This is onset detection by energy considerations, and the following definitions apply:

$A(t)$ the train of displacements due to the potential P-energy

τ_α the time period over which the integration of the displacements takes place

$B(t)$ also the train of displacements by the potential P-energy

τ_β the time period over which this latter integration takes place.

We note that $A(t) \supset B(t)$ and that $\tau_\alpha < \tau_\beta$.

If, then, the ratio "γ" exceeds a prescribed threshold $\left(\geq Unity\right)$, an onset is deemed to have occurred.

In both cases, when the ratio γ is exceeded, then a marker is placed against the lead sample belonging to the short time segment. The recording of such markers is achieved in the code, which handles input and output to the files 01 and 06, as described at Annex A.

The following diagrams represent the output just referred to. (The scale representing the time indexing on the lower diagrams is half that of the upper diagrams in each pair).

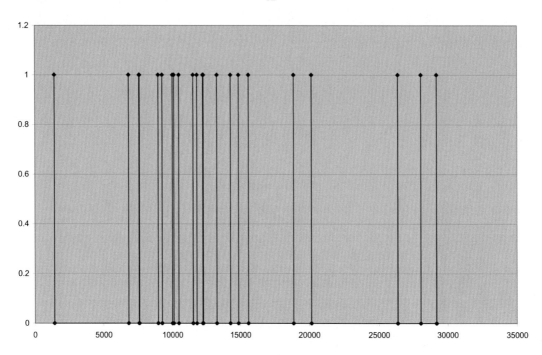

Figure 3.1 Onset Marking

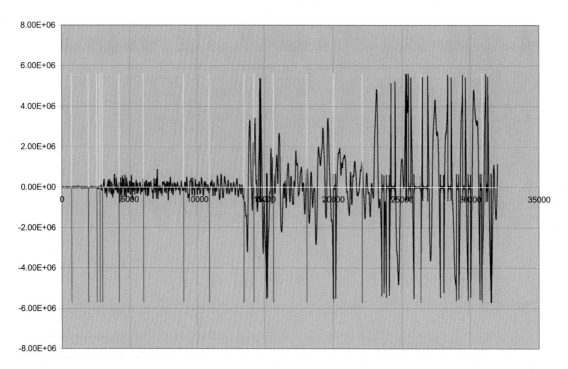

Figure 3.2 Onset Marking placed within Seimogram

The results are held in file 06 (see Annex A).

Further Methods for Detecting Energy Onsets in Seismogram Records

Here, we demonstrate a means for identifying the positions of fresh onsets of energy within a *recorded* seismogram, although the structure of the method would also lend itself to a real-time operating mode. This technique is formed from a dual test followed by a consolidation phase. The first test uses a version of the statistical t-test adapted for comparing means when their corresponding variances are, or may be, non-homogeneous. The second is the comparison, by ratio, of power levels, or means, between long- and short-running buffers. Both tests have to be in agreement in identifying the inception of an energy onset. The first triggers the second. Each of the identifications will produce a sheaf of alerts as the small buffer moves forward across the onset zone. These sheaves are consolidated in a subsequent process by means of a semblance technique.

The seismogram used in this trial is one taken from those generated by an earthquake in the Kuril Islands, stated to be on 19 April 2013.

$$46.224N;\ 150.783E$$
$$H_d\ \left(Depth\right) = 112.20km$$
$$MW\ 7.3$$

This information was accessed by a beta version of Wilber 3 (by courtesy of IRIS).

This seismogram was integrated to displacement level and band filtered in the band (0.1, 10.0) Hz. The latitudinal displacement for this recording seismograph was 40.10^0 away from the epicentre as pole.

The subsystem tried here is to determine the onset points of energy such as a P wave, S wave, Love wave, or Rayleigh wave, in an incoming seismogram. It comprises two phases:

1. The initial detection tests
2. A consolidation phase by a semblance function method

The initial detection consists of a pair of tests, which must both agree on the possible existence of an onset. The second element in this pair is a ratio of power integrals or means, while the first test, or trigger, is a t-test usually used in the presence of non-homogeneous variances.

T-Test with Non-homogeneous Variances

It is possible that the required condition for a t-test proper—the homogeneity of variances—is not fulfilled in the context of the disturbances due to the onset of varying types of energy (Neville AM, Kennedy JB, "Basic Statistical Methods for Engineers and Scientists", Intertext Books, 1964). Thus, the variance estimates probably cannot be pooled. The ratio of the standard deviation of the two samples means (using absolute values from the small and large buffers respectively)

$$\theta = \tan^{-1}\left(\frac{s_{\bar{x}_1}}{s_{\bar{x}_2}}\right)$$

and the statistic

$$\frac{\overline{x}_1 - \overline{x}_2}{\sqrt{s_{\overline{x}_1}^2 + s_{\overline{x}_2}^2}}$$

are formed. If the latter is found greater than a value, d, (parameterised in tables by θ, $v_1 = (n_1 - 1)$, and $v_2 = (n_2 - 1)$, then the difference between the means of the small and large travelling buffers (of the absolute values taken in each sample) would be significant at a specific probability level. The power level comparison test would then be triggered. Here, n_1 and n_2 are sample sizes. And v_1 and v_2 are the corresponding degrees of freedom, for the small and large buffers respectively.

The value, d, is derived from a table in *Basic Statistical Method for Engineers and Scientists* (table A-9). This same table is cited in a work by R. A. Fisher and F. Yates, *Statistical Tables for Biological, Agricultural and Medical Research*. This value is derived by interpolation in a three-space (θ, v_1, v_2).

In the current trials, $n_1 = 14$ and $n_2 = 600$, and the level of significance, at which d was interpolated, was the 0.01 probability level.

The Power Comparison

The power comparison has been touched upon above and can be attempted in two, if not three, different ways:

1. Applying an FFT to the seismogram to determine the power spectra PSD of the two time segments of the detection scheme
2. Integrating the absolute values of the seismogram sample data
3. Also, integrating the square of such values to estimate the mean power values for the two time segments forming the detection mechanism (3)

Having performed the necessary integrations, then we might say

$$\gamma = \frac{\alpha_P}{\beta_P},$$

where α_P and β_P are the respective results of integrating over the short and long buffers.

In all cases, when the ratio γ exceeds a designated threshold, ($\geq Unity$), this confirms the finding. A marker is placed against the trailing sample belonging to the short time segment (small buffer). The recording of such markers is achieved in the code, which handles input and output to the data structures, and files, which communicate with the routines responsible for locating the epicentre and Hypocentre of the event.

Figures 3.3 and 3.4 show the effect of applying method 2 and method 3, respectively, at complementary threshold levels of the initial t-test.

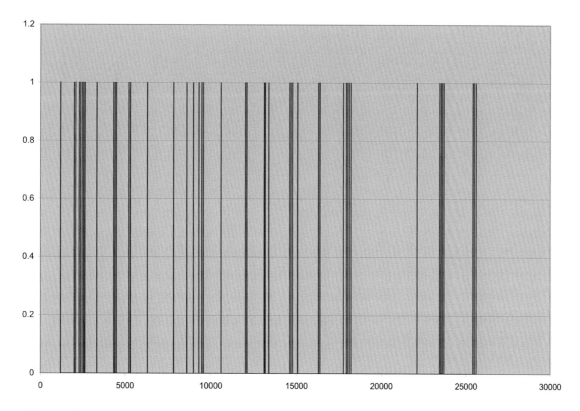

Figure 3.3. Energy onset markers generated by method 2 at 0.025 probability level

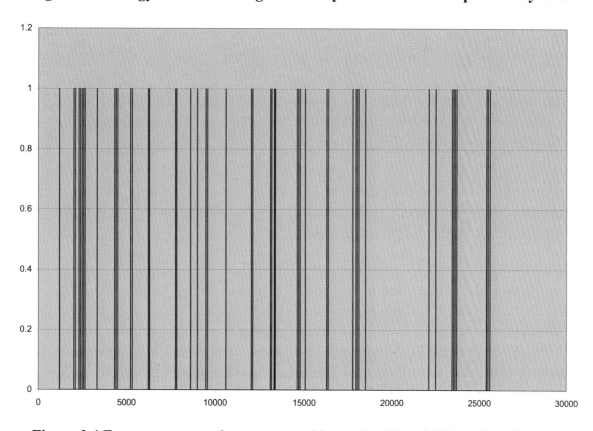

Figure 3.4 Energy onset markers generated by method 3 at 0.005 probability level

Consolidation by Semblance Function

The diagrams in figures 3.1 and 3.2 represent the superposition of markers on the object seismogram trace, as described above.

This marker superposition takes place as a two-phase process:

- First, the individual traces are scanned in tandem by a t-test between the means of the small and large buffers, which may then trigger detection of an energy change as a power-integrative process or a mean ratio process. A set of possible energy arrivals is collected for each trace.
- Second, the set of energy arrivals are scanned for coherent bunches of possible arrivals. Where such coherence is found, the leading edge of the bunch is marked and superimposed upon the appropriate seismogram as a white (in the present case) vertical line. The means used for scanning for coherency is a choice of four forms of semblance function, which compare vectors comprising the actual distances of each energy change from a given and moving base with a standard vector representing the close-packed and banded state b or ϕ_i in the four forms below). The comparison comprises the formation of an inner product between these two types of vectors. Thus, total compliance is unity, and total disparity is zero. The four forms are (using the notation found in [Fomel S, "Velocity Analysis using AB Semblance", Geophysical Prospecting, Vol 57, pp 311-321, 1960.]):

  o $\gamma(a,b) = \dfrac{a \cdot b}{|a||b|}$

  o $\beta(a) = \gamma(a,c);$, where c is a sequence of constants

  o $\alpha(a)$, where the trend $b_i = A + B \cdot \phi_i$ has been substituted into $\gamma(a,b)$, above

  o $\alpha(a) = \sqrt{Unity - \dfrac{\left(\sum\limits_{i=1}^{N}(a_i - b_i)\right)^2}{\sum\limits_{i=1}^{N} a_i^2}};$

- The coefficients γ, β and α are subjected to a test of significance (with a choice of 5 per cent or 1 per cent level, in an equal tails test) using correlation coefficient tables quoted by (Neville AM, Kennedy JB, "Basic Statistical Methods for Engineers and Scientists", Intertext Books, 1964) and taken from (Fisher RA, Yates F, "Statistical Tables for Biological, Agricultural, and Medical Research, Oliver & Boyd, Edinburgh, 1963).

The detected onsets are displayed as a superposition on the traces since, although the routines that perform these functions work automatically and independently (once their optimal guiding parameters have been ascertained), their values should be checked by the human eye. Those which are likely to correspond to P-wave, S-wave, Love-wave, and Rayleigh-wave onsets (in time) should be passed on to those routines responsible for locating the epicentre and the hypocentre of the given event. In fact, an accurate definition of the position (time-axis index) and timing properties of each line is passed on from tables held within the system.

Given the above information on the selection of the points of energy change within the integrated seismograms, we add further description on the way in which such a process can be deployed.

In the current implementation in which this method is under test, a pilot seismogram is chosen. Parameters for the reduction process, using the semblance function, are chosen (in other words, length of vectors and the acceptance threshold, 0.025 or 0.005 probability level). The result of using these chosen parameters is displayed and viewed. On viewing, the user has the option to reselect the parameters and try the process again. This cycle may be repeated until it is judged that the parameter selection is optimal. This final parameter selection is then allowed to run over the entire set of integrated seismograms currently under consideration.

Conclusions on Method

Initially, the threshold level for the comparison test (which was a ratio of power integrals or means) was simply set to *Unity*. Subsequently, it was thought to replace this simple power threshold (see under **PWTST** section at Annex A) with a ratio of means, in the following manner:

$$\frac{\overline{x}_1}{\overline{x}_2} > \left(Unity + \delta \cdot \left(\frac{s_2}{\overline{x}_2} \right) \right)$$

Here, s_2 represents the standard deviation in the current large travelling buffer (in this case of 600 samples). And δ is the fraction of the standard deviation, above which level the test is considered to have detected an onset.

The values of δ were chosen to be 1.0, 1.96, and 2.576. These represent probability levels of 0.16, 0.025, and 0.005 respectively in single tails. The corresponding semblance parameters chosen to process the markers generated by this new combined first phase of the testing were length 8, 0.05 significance; length 5. 0.05 significance; and length 3, 0.05 significance. In all cases, semblance type 3 (above) was chosen.

The functional relationships between the set of parameters that define this suggested procedure, namely the detection of energy onset within incoming seismograms, need further examination and analysis with a view to determining their optimal setting in dynamic (real-time) or static (recorded) circumstances.

These parameters are:

1. Length of small travelling buffer
2. Length of large travelling buffer
3. Value for δ, above
4. Semblance parameter length (number of degrees of freedom)
5. Semblance parameter significance (level at which the test signals an onset
6. Significance level for t-test

Another consideration that presents itself is choice of the method used for the second test in the initial phase. The three options include:

1. Power spectra (see under **PWTST** at Annex A)
2. Ratio of power integrals (see under **PWTST** at Annex A)
3. Ratio of means

The following pairs of figures—3.5 and 3.6, 3.7 and 3.8, and 3.9 and 3.10—show the results of some trials using the parameters defined above.

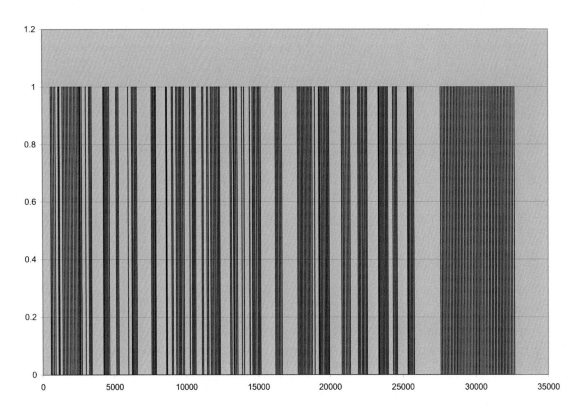

Figure 3.5. Comparison level at 0.16 probability level $1.0 * \sigma$)

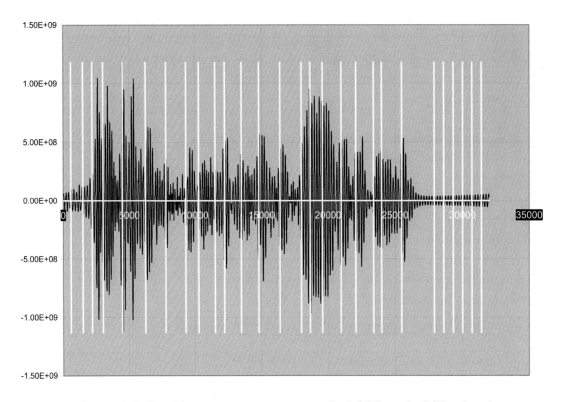

Figure 3.6. Semblance parameters: nu=8; 0.005 probability level

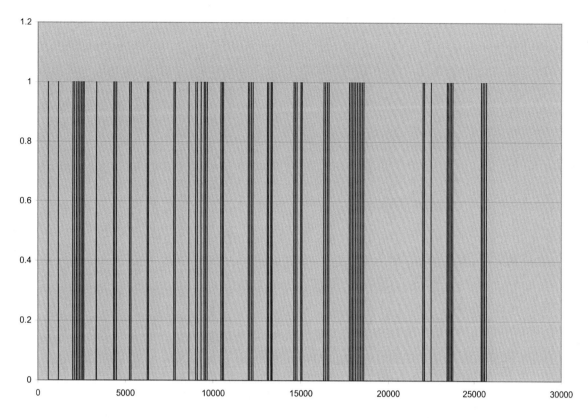

Figure 3.7. Comparison level at 0.025 probability level $(1.96 * \sigma)$

Figure 3.8. Semblance parameters: nu=5; 0.005 probability level

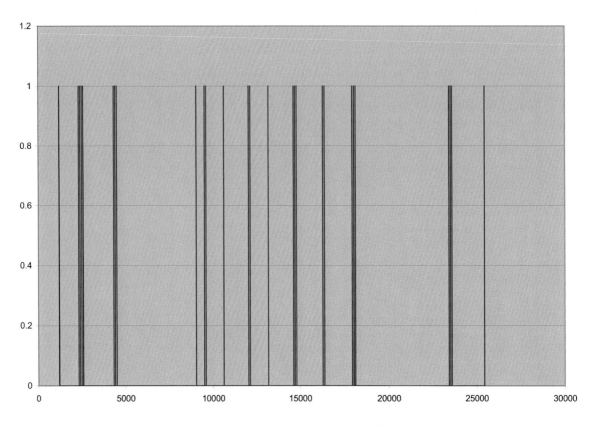

Figure 3.9. Comparison level at 0.005 probability level $(2.576 * \sigma)$

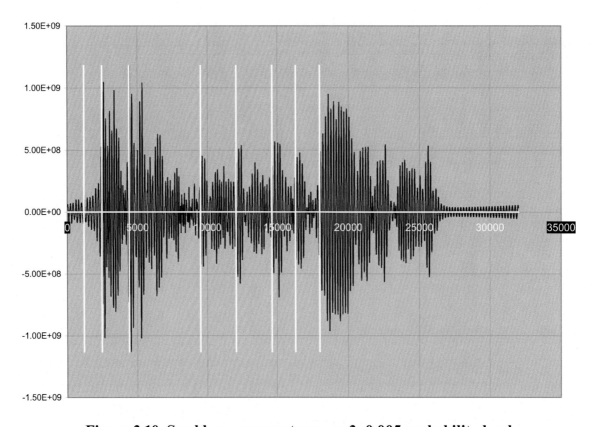

Figure 3.10. Semblance parameters: nu=3; 0.005 probability level

Estimation of Epicentral Coordinates from Time Differences within Individual Seismograms

For two separate arrivals at a given point in space, originating from the same source and having travelled with different velocities, we can write

$$v_0 t = v_1 \left(t + \Delta t \right)$$

since each has passed over the same distance. Here, $v_0 t$ corresponds to the distance covered by that arrival with the greater velocity. $v_1 \left(t + \Delta t \right)$ is the distance covered by the arrival with the lesser velocity.

And these two distances are the same.

The time to origin for the first arrival would be

$$t = \frac{v_1}{v_0 - v_1} \cdot \Delta t$$

Thus, knowing Δt, the time difference between the two arrivals, by observations on the given seismogram, we may write

$$s = v_0 t = \frac{v_0 v_1}{v_0 - v_1} \cdot \Delta t$$

or

$$s = v_1 \left(t + \Delta t \right) = \left(\frac{v_1}{v_0 - v_1} + Unity \right) \cdot v_1 \Delta t$$

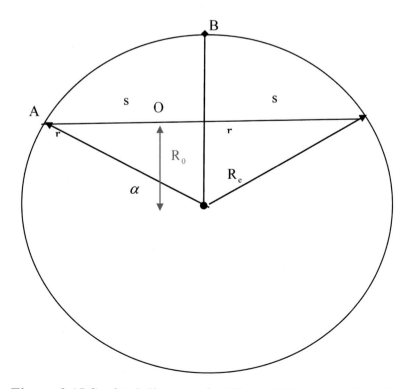

Figure 3.15 Sagittal diagram for Time-difference estimation

31

and s is the distance travelled from the source event by both wave species. From this, we can construct the above diagram. A and B are positions of the source and the receiving station respectively. R_e is an effective earth radius. In this sagittal diagram, we find

$$\alpha R_e = s \;\Rightarrow\; \alpha = \frac{s}{R_e}$$

$$r = R_e \sin\alpha; \quad R_0 = R_e \cos\alpha$$

In three-space, the coordinates of the point O, in the same frame as the coordinates of point B, are

$$a_0 = \frac{R_0}{R_e}\cdot a; \quad b_0 = \frac{R_0}{R_e}\cdot b; \quad c_0 = \frac{R_0}{R_e}\cdot c,$$

where the position of O is (a_0, b_0, c_0), and the position of B (the station) is (a, b, c). The point O, therefore is the centre of a sphere:

$$(x-a_0)^2 + (y-b_0)^2 + (z-c_0)^2 = r^2$$

The surface will include the source position at A. Thus, for many stations and correspondingly observable Δt_i we get a set:

$$\{\alpha_i\} = \left\{\frac{s_i}{R_e}\right\}$$

From this set, we might generate a system consisting of n spheres:

$$(x-a_i)^2 + (y-b_i)^2 + (z-c_i)^2 = r_i^2; \quad i = 0,(n-1)$$

The mutual intersection of these spheres (in the absence of error) can be thought to represent the position of epicentral coordinates for the source at A. This intersection is calculated as a linearised least-squares problem in program DLTST 00.

A program was specifically written to demonstrate this concept:

EpiCentreByTimeDifference.cpp

A direct minimisation of the least-squares cost function,has also been added to the algorithm depicted in this and implemented in DLTST 00.

$$S_n = \sum_{i=0}^{i=n-1}\left(\sqrt{u_i}-r_i\right)^2; \quad u_i = (x-a_i)^2 + (y-b_i)^2 + (z-c_i)^2$$

This uses the output from the linearised version of the last system, above, as a starting value for its iterations. This is a Gauss–Newton iterative algorithm, which will, firstly, use Jacobean matrices but will pull in a Hessian matrix if the main iterations are going astray.

When both these methods have provided their respective solution vectors, in terms of two estimates of the epicentral coordinates, the user is free to choose which estimate to accept as a basis for further processing.

The routine DPTST 00 acts in exactly the same manner as DLTST 00, except that the curved spherical surface is now assumed, in some sense, to be flat. The stations are all coplanar, thus giving a solution in the plane if it is assumed, further, that the equations of a sphere can be used to locate the possible epicentre lying in this plane.

A further routine DQTST 00 attempts to estimate the epicentral position by considering those sensors that would appear to lie within about nine degrees of distance from the event. This it does by:

- Projecting the relevant stations onto a plane, which is the plane of "closest approach" to this cluster of stations
- Using the calculated radii from given time difference pairs to calculate a position on this plane, as if the radii and the station positions were inscribed into this plane

This position is projected back onto the sphere and is taken as an estimate for an epicentral location.

Direct Estimation

Using slightly different notation from that used at the outset, we can say that the development for the simple spherical shell or lamina is given here as a conceptual equation system—in particular:

$$\underline{x} \cdot \underline{a_i} = R^2 \cos\left(\frac{\overline{c}_0\left(t_i - \Delta t\right)}{R}\right); \quad i = 0, (n-1)$$

where:

$\underline{x}$ is the location of the Seismic Emission in cartesians.

$\underline{a_i}$ is the location of the ith sensor in cartesians.

R is the radius of a spherical shell.

t_i is the ith arrival time relative to the lead sensor time.

Δt is the time-to-origin, i.e. the arrival time at the lead sensor.

$\overline{c}_0$ is, in some sense, an averaged wave propagation velocity.

Then the solution vector for the above equation is

$$\underline{s} = \begin{bmatrix} \underline{x} \\ \Delta t \\ \overline{c}_0 \end{bmatrix}$$

The Cartesians referred to here are coordinates within a space frame whose origin coincides with the centre of the object sphere. Thus:

$$\|\underline{x}\| \approx R; \quad \|\underline{a_i}\| \approx R.$$

This fact (or supposition) allows the following derivation:

An arc length is given by an expression such as $R\theta$, where R is the radius of the sphere or circle in question. Therefore, considering the sphere on whose surface emissions are being transmitted and removing the coordinate system origin to the centre of this sphere we get

$$R\cos^{-1}\left(\frac{xa_i + yb_i + zc_i}{\|\underline{x}\| \cdot \|\underline{a_i}\|}\right) = \overline{c}_0\left(t_i - \Delta t\right); \quad \iota \in \left[0, (n-1)\right],$$

But :

$$\|\underline{x}\| \cdot \|\underline{a_i}\| = R^2.$$

So :

$$R\cos^{-1}\left(\frac{xa_i + yb_i + zc_i}{R^2}\right) = \overline{c}_0\left(t_i - \Delta t\right).$$

Transposing functions gives

$$xa_i + yb_i + zc_i = R^2\cos\left(\frac{\overline{c}_0\left(t_i - \Delta t\right)}{R}\right); \quad i \in \left[0, (n-1)\right],$$

as above. This system is then solved for $\left(x, y, z, \Delta t, \overline{c}_0\right)$.

To fix $\underline{x}$, using this system, $n \geq 5$. It is notable that the above system is in the form of a set of planes, the lengths of whose normals to the origin are time dependent and oscillate as cosine functions. We can write, therefore, for one such plane

$$\underline{a_i} \cdot \underline{x} = f(t_i)$$

$$f(t_i) = R\cos\left(\frac{\overline{c}_0}{R}(t_i - \Delta t)\right)$$

$$f(t_i) \in \left[-R, R\right]$$

Here, $\underline{a}$ is now a vector of *direction cosines* for the ith normal. And $\underline{x}$ is the Cartesian point of emission in the space frame within which the Sphere is embedded and whose origin coincides with the centre of the sphere.

These direction cosines will be given by

$$\underline{a_i} = \frac{(a,b,c)_i}{R}$$

A wave front corresponding to a particular wave velocity, $\bar{c}_0$, (in some sense a mean value for a medium that has been assumed to be isotropic) is defined by the intersection of the oscillating plane and a sphere of radius R.

It is also of note that the period of oscillation for the entire set of planes is given by

$$T = \frac{2\pi R}{\bar{c}_0}$$

The restitution of $\underline{x}$ can be considered to represent the point where all n planes mutually intersect and where indeed the point of intersection lies on the spherical surface of radius R.

The solution of system (1.7) proceeds in two stages. An initial approximation to the solution vector is found by a scanning method using a "spider" moving over a region of a surface defined as a time/velocity space (see Annex J).

Using this initial value, a Gauss–Newton "descent" method is employed to find the root of system (1.7), considered as a highly non-linear least-squares cost function.

These methods are implemented in both

DynamicSphericalDeterminationAvto

and its companion

StaticSphericalDeterminationAccelerated.

We can give more consideration to this direct method in Chapter 4.

Testing for "Oceanic" Epicentres

An epicentre is determined as "oceanic" by using a structure made up of numerous and variably spaced lunes covering the terrestrial globe, from north to south. Rapid access to each lune is achieved using the value of the epicentral longitude. Each lune possesses a set of intervals that define its intersection by some corresponding set of land masses. These intervals are defined by latitude so that, if the epicentral latitude is not found within any such interval, an oceanic epicentre is inferred. (These intervals are adjusted for coastal proximity).

Scanning for Hypocentre

The topic is resumed from chapter 5 onwards, but at this juncture we give no discussion of point-to-point ray tracing and the "interpolative tabular scan" (which is adopted here). Nor, indeed, do we discuss of the general inversion problem and simultaneous inversion, as expounded initially by W. H. K. Lee and S. W. Stewart, having been developed from Geiger's method.

Fault Plane Determination

This would follow on from the determination of the hypocentral depth (H_d), with nodal planes possibly being established initially using the focal sphere arrival polarities projected onto a unit circle. Three-dimensional interactive visualisations of the possible dip, strike, and slide can then be made available.

The Bathymetric Mesh

This mesh would be difficult to set up in continuity for the resulting polar regions on the terrestrial sphere however this was tilted and/or rotated. Instead, such regions could be approached to within ten or twenty degrees latitudinally. The mesh could be predicated on a grid of appropriate intervals of latitude and of longitude the local earth radius (international ellipsoid) could be taken into account.

Tsunami Waves on a Sphere

This subsystem would attempt to predict the progression of a tsunami wave in an appropriately reduced target oceanic region.

The non-dispersive and dispersive phases of the development would be modelled. So too would deductions concerning landfall information (first arrival time, height of incoming mareograms and likely inundation, and the influence of the continental shelving).

Certain assumptions will be made concerning the generating mechanism. This will interact with any knowledge gained in the fault plane determination (given above).

Gravitational anomalies can be taken into account, together with the above bathymetric mesh. This mesh can be loaded in fine definition from disc to cover the target oceanic region chosen at the outset of the modelling procedure.

Intelligenced Software for the Possibility of Sensing Event Precursor Patterns

The many types of information schemas that this prototype system would work with are mooted in, for example, by Koyama J.[10]. Each such schema has a proven applicability here. It appears that, when schemas are coupled with known processes working in the background environment to those locations where seismic events can be observed (and, indeed, tracked in real time), realistic fits, modelling these seismic phenomena, are produced. This applies to swarms, foreshock sequences, and aftershock sequences.

An expert or intelligent system working in real time might be deployed to optimally select those schemas that were following the evolution of such sequences most closely.

Therefore, when a schema is indicated as critical or approaching criticality, it may be possible for the system to send notification of this circumstance. It is expected that the database levels of possible schemas, together with those of their coupled background processes, would evolve dynamically as real-time experience accumulated.

---

[10]   J Koyama, *The Complex faulting Process of Earthquakes,*,Springer, 1997.

Chapter 4

Hypocentre Location Using Tabular Data
with Preliminary Illustrative Results

This chapter attempts to describe, with illustrative examples of its output, a small subsystem intended to form part of a body of intelligenced seismic software. This software is intended primarily:

1. To perform rapid determination of hypocentres, once appropriate epicentres are known
2. To form a simulation system to assess such a process
3. In so doing, to test models of earth velocity structure (for example, PREM, iasp91, and ak135), as well as the efficacy of various point-to-point ray-tracing techniques

In effect, this system is a table-driven scan and works on tabular data structures set up by combining an appropriate point-to-point ray tracer with any particular velocity structure proposed for earth's interior.

Structure of Hypocentre Scanning Subsystem

The procedure for finding hypocentre foci, using tabulated timing information, requires first the epicentral coordinates of the given event. Given this, procedures—both proximal and teleseismic—can be put into place. At the time of this writing, these procedures will locate to a depth of 1,000 kilometres within a time of less than 0.25 seconds (using a granularity of 2,000 scanning points on a 3.2 gigahertz Dell XPS600 machine).

The overall diagrams for the sub-system that achieves this are depicted as **Figure 4.1** and **4.2** below.

Data Entities are defined by brackets in red. Programs (active nodes) are defined by brackets in black. The data flow is indicated by arrows.

Figure A01 describes the data flow in scanning for a hypocentre depth, once the epicentre is known.

The entities 00AA, 00BB, 00CC, and 00DD represent the tripartite elements referred to in the text. They are formed from the entities 00, 01, and 02, which contain the timing, take-off angle, and accuracy data (the latter not currently being utilised in the present state of implementation) for a specific run of the table-generating nodes 01 to 02 (where we refer as well to Figure 4.2).

These nodes are named

"TimeBased01.cpp"; "TimeBased02.cpp"

and are described in the main text. Node 03 identifies the routines associated with the program ETTST 04 (05).cpp, and node 04 identifies the routine ETTST 01.cpp. All results are currently viewed by communication with Microsoft Excel.

The entity sequence 00AA to 00DD (figure A01) represents tripartite elements formed by the combinations of ("radial"/ak135), ("radial"/PREM), ("eikonal"/ak135), ("eikonal"/PREM), relating the ray-tracing method to the velocity structure to be used.

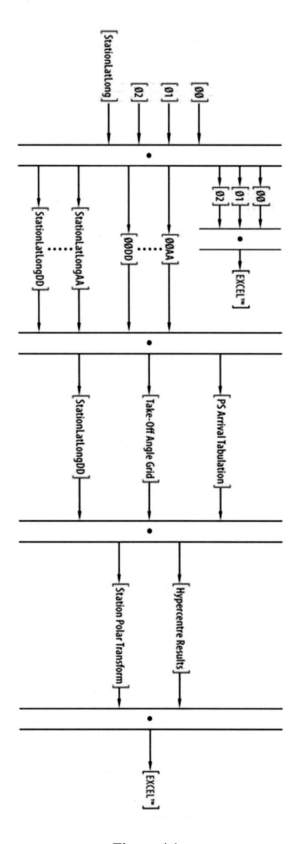

Figure 4.1

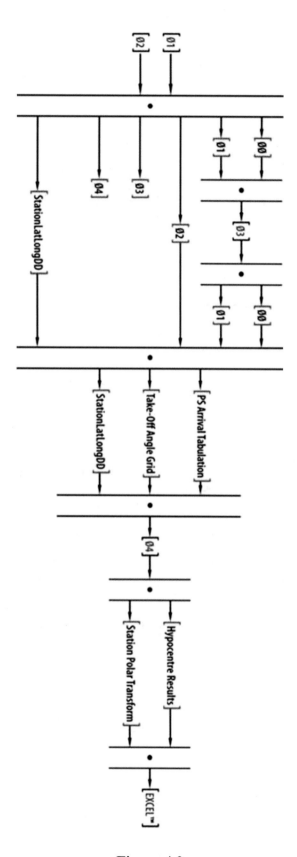

Figure 4.2

Currently, the transition of any tripartite element (entity) to the three main input elements for node 04 (see Figure 4.2) is by the separate nodes MSTST.cpp, MXTST.cpp and/or MQTST.cpp.

The nodes mentioned in the main text as

"TimeBasedSingleRay.cpp"

and

"TimeBased03"
"TimeBased03BB"; "TimeBased03LL"
"TimeBased03QQ"; "TimeBased03SS"

with

"TimeBased04"

(being stand-alone test routines, primarily written to determine the levels of self-noise in the system) are not included in the above diagrams. These latter deal with the finished production form of this subsystem.

The method here is an interpolative scan. To this end, so far, four interpolation schemes have been implemented:

1. Linear, linear
2. Linear, cubic
3. Cubic, cubic
4. Lagrange, Lagrange

The initial method given within each pair refers to interpolating in depth, while the second refers to interpolating in latitudinal distance (or displacement).

During the generation phase, the tabular data is currently formed into three separate tables:

1. Travel times
2. Take-off angles
3. Accuracy of the point-to-point (P2P), ray path generated

Each of these variables is tabulated in terms of depth and latitudinal displacement.
These tables are generated by either of the program nodes:

"TimeBased01.cpp"
"TimeBased02.cpp"

which make use of routines to be found in

"EikonalScanningTools.h"
"HypocentreScanningTimeBased.h"

The first header file contains files associated with an eikonal equation-based ray trace system. The second contains files associated with a radial and a Lagrangian ray tracing system. The bases for these systems are described in chapter 5.

The *possibility* of locating hypocentres very rapidly using stations from within a diameter of nine degrees, and so *proximal to the epicentre*, must also be considered here. The routines required for this are detailed and described in chapter 5. They are Lagrangian and are to be found in the file called

"HypocentreScanningTransBased.h"

Ray Tracer Structure

Each P2P ray tracing system consists of a three-tier structure beneath a zeroth, or root-activating, level:

1. P2P controller
2. Parameterisation interface
3. Ray tracer (radial, eikonal, or Lagrangian)

The third element generates the data defining:

- Length of Ray
- Travel-time of Ray
- Accuracy to Target-point in meeting Latitudinal Distance and in terminating at given Earth Radius (currently taken to be 6371.0 km)
- Take-off angle of Ray
- Track of Ray (coordinates of each ray-segment)

The *second element* 2., above, takes the P2P path-defining parameters and transforms to 3-space unit tangent vector, which defines the take-off requirements for use by the *third element*.

The *first element*, 1., above, contains the logic that ensures that the required P2P path is generated. This logic consists of a binary interval search, which can be described by the following: The steering loop scans for a hit (in other words, a bracketing) on the target point (a relative latitudinal displacement) using a "binary chop" procedure.

In essence this process proceeds as follows. A grid of N $(N = 100,$ say$)$ equally spaced points is dynamically laid down within an angular range, $[Zero, 180]$. The required target latitude is known. This grid of angles is taken to be a set of take-off angles and so is used to form overlapping pairs as

$$[P_0, P_1]; [P_1, P_2]; \ldots; [P_{N-1}, P_N]$$

$(P_0$ is *Zero* degrees, while P_N is 180$)$

When a pair of such take-off angles is found to generate a corresponding pair of resultant latitudes that bracket the target latitude, then that range of the take-off pair is divided into *two equal parts* and is scanned using new values for its bounds taken from the original $[P_i, P_{i+1}]$:

$$\left[P_i, \frac{P_i + P_{i+1}}{2}\right]; \left[\frac{P_i + P_{i+1}}{2}, P_{i+1}\right]$$

The scan is *repeated using such duplet intervals* until the bracketing around the target latitude, or the size of the generating duplet, has been constricted to within a predetermined accuracy threshold.

In all, the two ".h" files contain all routines needed to implement the architecture of these 3-tiered systems, which are proposed to work to some extent teleseismically.

However, in contrast, the approach that uses the closely lying proximal stations and assumes flat stratification, mentioned above, uses a Lagrangian approach to effect the control of the P2P tracers employed in this latter context. This control technique is described in chapter 5.

Ray Tracer Action

There is a second aspect to the P2P ray tracing capability. This is the visualisation of a single ray path or set of ray paths. This is controlled by a program node:

"TimeBasedSingleRay.cpp"

This program routine uses the P2P three-tier stack, starting from the second level, and supplies only:

1. Earth velocity model (for example, ak135, PREM)
2. Depth of source
3. Take-off vector
4. Granularity of ray elements

It does so without the target-point constraint, which is implemented by the first layer of these stacks. Three examples of ray trace groups are given at Annex F.

Building Tabular Data

The subsystem for table building, briefly described so far, acts to generate a data bank of quadri-partite elements. Each element is a function of:

1. Table spread parameters in depth and longitudinal displacement
2. The parameters defining the ray constructions
3. The type of (currently radial) earth velocity model used

Normally, each of the three tabular components forming any one of the banks of elements is dimensioned from 101×101 down to 51×51. The four components are:

1. Travel times
2. Take-off angles
3. Ray Path length
4. Accuracy information

Such a four-part element can take up to six and a half hours to prepare, depending on how fine the granularity and which technique is used for the ray tracing.

There are also subsidiary vector tables giving the structure of the defining depth and latitudinal displacement axes.

At this time, a manual interface serves to select and transmit any particular tripartite data element from its position in the data bank to the active input structure that serves the hypocentre location scan.

Any error in these potential database elements is suppressed by running the program routine:

"ETTST 04.cpp"

This routine will interpolate the results (output) of "TimeBased01.cpp" at exactly the required tabulating latitudinal points. After this, the four-part element assumes its position within the collection of tables that form the database. An example of a set of data elements finished in this way is given in Annex F. In this case, the elements are formed from both the output of the eikonal ray tracer and the radial tracer. The interpolation within "ETTST 04.cpp" is linear and also by the method of Lagrange.

Action of Hypocentre Scan

The hypocentral scan can use all four interpolation regimes given above in parallel or in tandem. It can also attempt to generate a consensus result from all four as a concluding value for the hypocentral depth, H_d. The main program node that achieves this is

"ETTST 01.cpp"

This is supported by routines within the file

"HypocentreScanningTimeBased.h"

given above.

The scanning method is basically made up from the considerations described in this section. Currently, three scanning methods are realised:

1. Direct reconstruction of a *relative set of arrival times* from tables for each depth point in the self-test, followed by matching to the sets of arrival times, which have also been structured relatively

and input (calculated by ray tracer) for the given (simulated) hypocentre foci. The self-tests are performed by the teleseismic scanners using the family of routines

"TimeBased03"

"TimeBased03BB"; "TimeBased03LL"

"TimeBased03QQ"; "TimeBased03SS"

and the proximal scanners, using the routine

"TimeBased04"

These generate a row of hypocentre foci. Each hypocentre is then located by a tabular scan. The foci located are plotted against the simulated foci, together with error performance indications. "BB", "LL", "QQ," and "SS" refer to types of the consolidation method, which is to be described below.

2. The second type is essentially the same as the first, but with a final consolidation by linear/linear (or cubic/cubic) reverse interpolation of the currently held location in the scan, which attempts to overcome the error induced by the finite (but hopefully small) step length of the tabular scan. This consolidation method proceeds in one of three possible modes, each of which concerns the manner in which the approximation to the absolute value of the raw arrival times is made.

3. The third is a direct look-up or interpolating scan of the given arrival times, taken as absolute durations, not relative values. This would be possible, having found the temporal and spatial coordinates of the corresponding epicentre. Any error in the timing of the epicentre will result in a bias. It is of note that, when the step length within the row of simulated hypocentre foci is the same as that for the tabular scan in the self-tests, then there is a strong possibility of achieving *zero error*. Even though neither of these step lengths may be "in sync" with the depth step length used for the structuring of the data tables used by the scan, this still remains a strong possibility.

The accuracy of the results produced by the scanning procedure can be increased by using a refinement, denoted here by the term *consolidation*.

This process can assume that a value for the epicentral time, (ε_t), for the event is given or has been ascertained prior to this hypocentral calculation being made. This is also necessary for the hypocentral scan to happen in the first place, since the relative positions of the participating stations must be calculated relative to the epicentral point so that the above tables can be deployed.

Let the residuals from the raw timing data be

$$\Delta s_i = \left(s_i + \delta s_i \right) - \frac{1}{n} \left(\sum s_i + \sum \delta s_i \right)$$

which is calculated by

$$\Delta s_i = T_i - \frac{1}{n} \sum T_i$$

where the T_i are the *raw arrival times* from the event picked within a universal time frame and where s_i is an interpolated timing from the data table of travel timings and is a function of the current depth assumed for the simulated hypocentre focus in the scanning process, and the relative latitude of the ith station. δs_i is the currently *unknown difference* between the interpolated timing (which is an *absolute value*) and that of the raw data absolute value.

(The interpolated timings are a collection of absolute values for the travel times of rays from depths to a set of stations at a collection of relative latitudes to the epicentral latitude, taken as a "polar value".)

Let the residuals for the interpolated timings be:

$$\Delta s_i' = s_i - \frac{1}{n}\sum s_i$$

Given the first set of residuals, Δs_i, then we can construct a *hypothetical set* of absolute values for the raw arrival times as

$$T_i = \Delta s_i + s_0$$

Here s_0 is a suitable reference value, (in other words, the mean value), taken from the set of modified raw arrival timings (which are now approximate absolute values). These timings are modified by the subtraction of the epicentral time, ε_t.

The difference between the two sets of residuals is:

$$\Delta s_i - \Delta s_i' = \left\{ \delta s_i - \frac{1}{n}\sum \delta s_i \right\}$$

We wish to find the absolute values for the raw timing data. This is

$$S_i = s_i + \delta s_i$$

but

$$\delta s_i = \Delta s_i - \Delta s_i' + \frac{1}{n}\sum \delta s_i$$

where

$$\frac{1}{n}\sum \delta s_i = \left(\overline{s_i + \delta s_i} \right) - \left(\overline{s_i} \right)$$

The modified raw data values above are assumed to approximate to their absolute values, with the observation that they will give a sharper minimum in the inner scanning process to identify that depth which most closely approaches the true value of the simulated hypocentre focus.

Given that this minimum most closely identifies the true absolute values of the raw data we can proceed as described below.

Since we have modified T_i as above, we now have $\left(T_i \approx S_i \right)$:

$$\left(\overline{s_i + \delta s_i} \right) \approx \frac{1}{n} \sum \left(T_i \right)$$

$$\left(\overline{s_i} \right) = \frac{1}{n} \sum s_i$$

Therefore, we can perform

$$S_i \approx s_i + \left\{ \frac{1}{n} \sum \delta s_i + \left(\Delta s_i - \Delta s_i' \right) \right\}$$

The values for S_i can now be used within the timings tables (absolute arrival times), from the set of data tables to reverse interpolate for a corrected estimate of the submitted hypocentral focal depth, if in the self-test regimen, or for an actual focal depth, if using real data.

A last variable in this group of procedures is the method or methods chosen for the interpolation processes. Currently there are three:

1. Linear
2. Collocating cubic
3. Lagrangian

These are used in all combinations for two-space interpolation, in terms of depth and latitudinal distance.

However, there remain several more interpolation options to be tested, both for accuracy and for speed. These include the methods of:

- Aitken
- Everett
- Bessell

Figures showing the results of a self-test with hypocentre foci from a depth of 1,000 kilometres to about 10 kilometres are appended. This test took place using cubic interpolation.

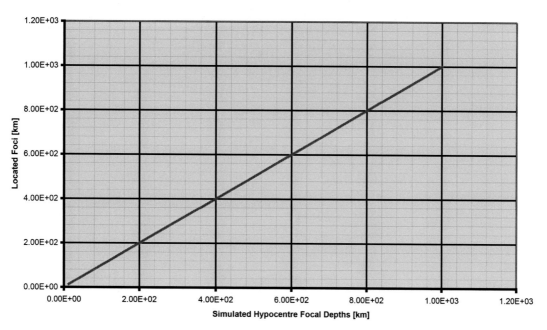

Figure 4.3

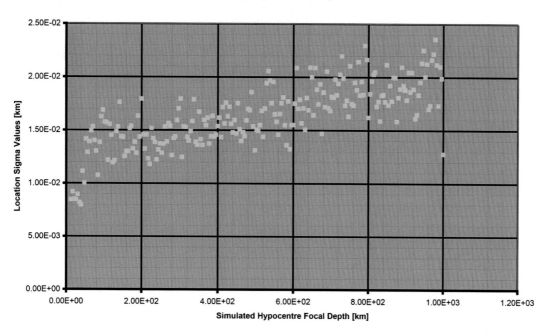

Figure 4.4

Hypocentre Location
Interpolating Scan
Pointwise Error

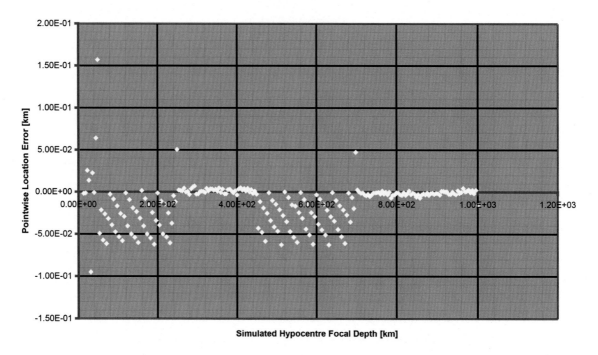

Simulated Hypocentre Focal Depth [km]

Figure 4.5

Hypocentral Location
Interpolating Scan
Accumulating Mean Error

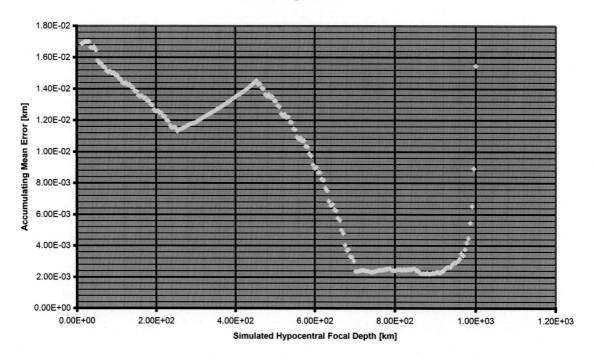

Simulated Hypocentral Focal Depth [km]

Figure 4.6

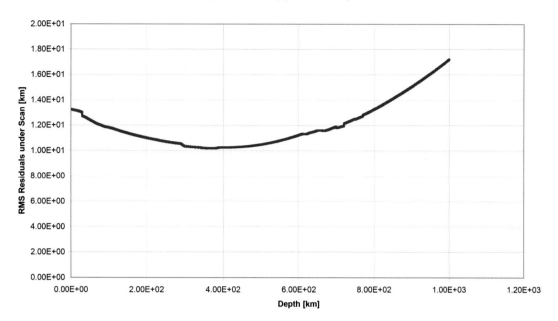

Figure 4.7

The construction of the set of *relative arrival times* follows the scheme described below.
The onset timings are made up as follows from the set T:

$$T \equiv \{t_j\}; \quad j = 0, (n-1)$$

Here, the t_j are elements of a universal time running on a continuous time base and represent arrivals of whatever wave species at stations $(j+1)$.
Thus,

$$\overline{t} = \frac{1}{n} \sum t_j$$

We generate a spread of deviations about this mean as

$$\tau_j = \{t_j - \overline{t}\}; \quad \text{j=0}, (n-1)$$

Similarly, using the point-to-point spherical ray tracing routines, from any position on the upward scanning trajectory, we can find a set of deviations i for $\{v_j\}_i$; $j = 0, (n-1)$.
i indexes a point in the scan, and j indexes the set of active seismic stations.
For each position in the scan we get

$$RMS_i = \sqrt{\frac{1}{n} \sum_j (\tau_j - v_{j,i})^2}$$

It is this set of values that is graphed against depth in Figure 4.7. The infimum is taken to correspond to a hypocentral depth, as a consensus from all four methods, in this case, of 377.25 kilometres. The value of 380 kilometres applies to the specific method (linear/linear) chosen for the purpose of drawing this trace.

Discussion

In order to exemplify the working of this system, a particular earthquake was selected for the IRIS database (Wilber 2). The chosen quake took place in the Sea of Okhotsk (neighbouring Sakhalin, Kamchatka) and was defined by these parameters:

Date: 14 August 2012
Magnitude: 7.7
Epicentre: Latitude = 49.78 N, Longitude =145.13 E
Depth: 625.80 km

The aim was to test the ability of current prototype software to locate the hypocentral focus of the earthquake, given an a priori value for the epicentral coordinates. See results in Annex H, which contains the input data to the scanning algorithm. Sensors from the IC, IU (GSN – IRIS/USGS), and II networks were used and the rotated coordinates of the active sensors chosen, were as follows:

Latitude; Longitude; Earth Radius; Latitudinal Distance

```
 1  7.8278122e+001  2.9021651e+002  6.3743410e+006  1.1721878e+001
 2  7.5084639e+001  2.0902961e+002  6.3743411e+006  1.4915361e+001
 3  7.5846494e+001  3.3667444e+002  6.3743410e+006  1.4153506e+001
 4  7.1903787e+001  3.0582870e+002  6.3743410e+006  1.8096213e+001
 5  6.5131356e+001  2.5965593e+002  6.3743410e+006  2.4868644e+001
 6  5.3717314e+001  1.2709943e+002  6.3743410e+006  3.6282686e+001
 7  5.3804387e+001  3.5957617e+002  6.3743411e+006  3.6195613e+001
 8  4.8352549e+001  3.0474889e+002  6.3743410e+006  4.1647451e+001
 9  4.8940511e+001  2.4213507e+002  6.3743410e+006  4.1059489e+001
10  4.5379893e+001  2.3567745e+002  6.3743410e+006  4.4620107e+001
11  4.5075146e+001  2.7568445e+002  6.3743411e+006  4.4924854e+001
12  4.2022641e+001  2.9316504e+002  6.3743410e+006  4.7977359e+001
17  3.7672423e+001  2.0267725e+002  6.3743410e+006  5.2327577e+001
18  3.4610069e+001  7.9626206e+001  6.3743411e+006  5.5389931e+001
19  3.1563612e+001  1.2332707e+002  6.3743411e+006  5.8436388e+001
20  3.0957597e+001  3.3014538e+002  6.3743410e+006  5.9042403e+001
21  7.7148675e+000  1.5361312e+002  6.3743410e+006  8.2285133e+001
22  8.9884930e+000  1.4155789e+002  6.3743411e+006  8.1011507e+001
23  7.9923221e+000  1.4848292e+002  6.3743411e+006  8.2007678e+001
24  8.8836059e+000  1.1641549e+002  6.3743410e+006  8.1116394e+001
25  1.0903190e+001  1.3811171e+002  6.3743410e+006  7.9096810e+001
26  1.5564476e+001  3.3547047e+002  6.3743410e+006  7.4435524e+001
27  2.1911219e+001  2.8532963e+002  6.3743410e+006  6.8088781e+001
28  2.4348868e+001  2.1870898e+002  6.3743411e+006  6.5651132e+001
```

(Latitudinal distance is the station distance from the epicentre in degrees. The earth radius given here is the radius used for restituting the station Cartesian coordinates in the earth space frame.)

This rotation of station coordinates takes place relative to the epicentre, which is considered a pole, itself being rotated to a latitude/longitude of ninety/zero degrees (latitude, co-latitude).

Results including:

$$H_d = 556.5 \text{ km [ak135]}$$
$$H_d = 861.875 \text{ km [PREM]}$$
$$H_d = 551.5 \text{ km [ak135]}$$

and

$$H_d = 863.376 \text{ km [PREM]}$$

are documented in Annex G. Again, these are somewhat wide of the quoted value of 625.80 kilometres. However, this run of the system is useful as an initial test with a small number of data inputs and has possibly achieved agreement to a fair order of magnitude.

Using relative P-wave arrival times and tables of P-wave arrivals against depth and latitudinal distance, the system scans for the depth of hypocentral focus and derives the take-off angles to each sensor (also from precalculated tables, as shown in Annex F).

In setting up all tables used in this trial, the earth velocity models "ak135", due to BLN Kennett et al. (1995), and a reduced version of PREM (Anderson and Dziewonsky, 1980) were used. The former model would have the potential to produce the more accuate results, while the latter would allow the tables to be generated more quickly.

Epicentre Estimation

Here, we also resume and develop a second type of equation for these purposes. Consider an arc length Γ on a great circle inscribed in a sphere. Consider also the chord it subtends of length $s = 2d$. Then, if we would wish to transform one into the other, in terms of length, to get Γ from s, we should multiply by the ratio

$$\rho = \frac{\alpha R_e}{2R_e \sin\left(\dfrac{\alpha}{2}\right)}$$

Thus,

$$\Gamma = s \cdot \frac{\alpha}{2\sin\left(\dfrac{\alpha}{2}\right)}$$

Here, α is the angle subtended by the arc and chord pair at the centre of the great circle, while R_e is the sphere (Earth) radius. To get the chord length from the arc length, we must multiply by the ratio

$$\rho_0 = \frac{1}{\rho} = \frac{2\sin\left(\dfrac{\alpha}{2}\right)}{\alpha}$$

Considering that we are dynamically observing events originating on or within a sphere and are aware of such events by virtue of the transmitted surface wave phenomena they emit, we can write, for the arc length Γ_i:

$$\Gamma_i = \overline{c}_0\left(t_i - \delta t\right)$$

Here:

$\overline{c}_0$ is the propagation velocity of the surface wave in question.

δt is the time to origin—in other words, the time taken for the energy to reach this sensor from the position of the event.

t_i is the elapsed time formed from subtracting the time at which the energy reached the lead sensor from the time it impinged upon the ith sensor.

t_0 is then, in fact, *Zero*.

Thus, if we have many sensors, we may have many arcs traversed in the like manner. Now we have

$$\frac{\alpha_i}{2} = \frac{\overline{c}_0\left(t_i - \delta t\right)}{2R_e}$$

Therefore, to convert the observed set of arcs, $\{\Gamma_i\}$, to chords, $\{s_i\}$, we can operate as follows

$$s_i = \Gamma_i \cdot \frac{2\sin\left(\dfrac{\alpha_i}{2}\right)}{\alpha_i} = \Gamma_i \cdot \frac{2\sin\left(\dfrac{\overline{c}_0\left(t_i - \delta t\right)}{2R_e}\right)}{\dfrac{\overline{c}_0\left(t_i - \delta t\right)}{R_e}}$$

substituting for gamma

$$s_i = \overline{c}_0\left(t_i - \delta t\right) \cdot \frac{2R_e\sin\left(\dfrac{\overline{c}_0\left(t_i - \delta t\right)}{2R_e}\right)}{\overline{c}_0\left(t - \delta t\right)}$$

so

$$s_i = 2R_e \sin\left(\frac{\overline{c}_0\left(t_i - \delta t\right)}{2R_e}\right)$$

If we then allow chords to be subtended between the sensor points on the surface of our sphere and the unknown position of the source, we can write the following system of equations for locating the approximate epicentre of the source:

$$\left(x - a_i\right)^2 + \left(y - b_i\right)^2 + \left(z - c_i\right)^2 = 4R_e^{\,2} \sin^2\left(\frac{\overline{c}_0\left(t_i - \delta t\right)}{2R_e}\right); \quad i = 0,\left(n-1\right).$$

With further development this system may be solved for the following:

$\underline{x} = \left(x, y, z\right)$, the position of source in Cartesians.
δt the time to origin, and
$\overline{c}_0$ the propagation velocity

We note that, for the solution of this system, we must also add a constraint. The Cartesians must lie on the surface of the sphere:

$$x^2 + y^2 + z^2 = R_e^{\,2}$$

We can, therefore, add or subtract this spherical locus from the above system, giving, say

$$2x^2 - 2xa_i + a_i^2 + 2y^2 - 2yb_i + b_i^2 + 2z^2 - 2zc_i + c_i^2 = R_e^{\,2}\left(4\sin^2\left(\frac{\overline{c}_0\left(t_i - \delta t\right)}{2R_e}\right) + Unity\right);$$

$$i = 0,\left(n-1\right)$$

Here, we have added. It is now possible to solve this system with great facility using the abbreviated form of Gauss–Newton, which springs from the quadratic approximation method. We have

$$-\underline{h}_i = \left(J^T J\right)^{-1} J^T \cdot \underline{f}\Big|_i$$

Here:
$\{\underline{h}_i\}$ represents the set of successive step lengths in the solution path
J is the Jacobean for the function vector $\underline{f}$, evaluated at the ith point in the iterations.

This is implemented in the node

"DynamicSphericalDeterminationAvto 04.cpp"

We must also remark that the spherical locus should be added to, or subtracted from, the system previously given in the main body for direct calculation of epicentre coordinates, namely

$$x \cdot a_i + y \cdot b_i + z \cdot c_i = R_e^{\,2} \cos\left(\frac{\overline{c}\,\left(t_i - \delta t\right)}{R_e} \right); \quad i = 0,(n-1).$$

This would lead to, on subtraction:

$$x\left(a_i - x\right) + y\left(b_i - y\right) + z\left(c_i - z\right) = R_e^{\,2}\left(\cos\left(\frac{\overline{c}_0\left(t_i - \delta t\right)}{R_e} \right) - Unity \right); \quad i = 0,(n-1).$$

This also can be solved with great facility using the Gauss–Newton method, as mentioned in the description above or as sketched in here. Or the full quadratic approximation method may be deployed:

$$\underline{x}_{i+1} = \underline{x}_i - \left(J^T J + \sum_{j=0}^{n-1} f_j H_{f_j} \right)^{-1} \cdot J^T \underline{f}\,\Big|_i$$

Here:

J is the Jacobean of the vector of functions $\underline{f}$ evaluated at iteration point i.
H_{f_j} is the Hessian of the function f_j evaluated at iteration point i.

This is implemented in the node

"DynamicSphericalDetermination Avto 03.cpp"

Chapter 5

The Ray Tracers Used

Eikonal Ray Trace Algorithm

This description follows, in part, that given by Lee and Stewart (Lee WHK and Stewart SW, "Principles and Applications of Microearthquake Networks", Academic Press, 1981). The three members of the ray equation system can be written (where n is the refractive index):

$$\frac{d}{ds}\left(n \cdot \frac{dx}{ds}\right) - \frac{\partial n}{\partial x} = Zero$$

$$\frac{d}{ds}\left(n \cdot \frac{dy}{ds}\right) - \frac{\partial n}{\partial y} = Zero$$

$$\frac{d}{ds}\left(n \cdot \frac{dz}{ds}\right) - \frac{\partial n}{\partial z} = Zero$$

Or, if divided by a reference velocity, v_0, this generates

$$\frac{d}{ds}\left(\frac{1}{v} \cdot \frac{dx}{ds}\right) - \frac{\partial}{\partial x}\left(\frac{1}{v}\right) = Zero$$

$$\frac{d}{ds}\left(\frac{1}{v} \cdot \frac{dy}{ds}\right) - \frac{\partial}{\partial y}\left(\frac{1}{v}\right) = Zero$$

$$\frac{d}{ds}\left(\frac{1}{v} \cdot \frac{dz}{ds}\right) - \frac{\partial}{\partial z}\left(\frac{1}{v}\right) = Zero$$

more concisely,

$$\frac{d}{ds}\left(u \cdot \left(\frac{dr}{ds}\right)\right) = \nabla\left(\frac{1}{v}\right)$$

Carrying out the differentiation,

$$\frac{du}{ds} \cdot \left(\frac{d\underline{r}}{ds} \right) + u \cdot \frac{d^2\underline{r}}{ds^2} = \nabla u$$

$$\frac{d^2\underline{r}}{ds^2} = \frac{1}{u}\left(\nabla u - \frac{du}{ds}\left(\frac{d\underline{r}}{ds} \right) \right)$$

$$\frac{du\left(\underline{r} \right)}{ds} = \frac{\partial u}{\partial x} \cdot \frac{dx}{ds} + \frac{\partial u}{\partial y} \cdot \frac{dy}{ds} + \frac{\partial u}{\partial z} \cdot \frac{dz}{ds}$$

$$\frac{d^2\underline{r}}{ds^2} = v\left(\nabla u - \left(u_x \cdot \frac{dx}{ds} + u_y \cdot \frac{dy}{ds} + u_z \cdot \frac{dz}{ds} \right) \cdot \frac{d\underline{r}}{ds} \right)$$

Disassociating,

$$\frac{d^2 x}{ds^2} = v\left(u_x - G \cdot \frac{dx}{ds} \right)$$

$$\frac{d^2 y}{ds^2} = v\left(u_y - G \cdot \frac{dy}{ds} \right)$$

$$\frac{d^2 z}{ds^2} = v\left(u_z - G \cdot \frac{dz}{ds} \right)$$

or

$$vu_x - v\frac{dx}{ds}\left(\frac{\partial u}{\partial x} \cdot \frac{dx}{ds} + \frac{\partial u}{\partial y} \cdot \frac{dy}{ds} + \frac{\partial u}{\partial z} \cdot \frac{dz}{ds} \right) = \frac{d^2 x}{ds^2}$$

$$vu_y - v\frac{dy}{ds}\left(\frac{\partial u}{\partial x} \cdot \frac{dx}{ds} + \frac{\partial u}{\partial y} \cdot \frac{dy}{ds} + \frac{\partial u}{\partial z} \cdot \frac{dz}{ds} \right) = \frac{d^2 y}{ds}$$

$$vu_z - v\frac{dz}{ds}\left(\frac{\partial u}{\partial x} \cdot \frac{dx}{ds} + \frac{\partial u}{\partial y} \cdot \frac{dy}{ds} + \frac{\partial u}{\partial z} \cdot \frac{dz}{ds} \right) = \frac{d^2 z}{ds^2},$$

where: $G = \frac{\partial u}{\partial x} \cdot \frac{dx}{ds} + \frac{\partial u}{\partial y} \cdot \frac{dy}{ds} + \frac{\partial u}{\partial z} \cdot \frac{dz}{ds}.$

Thus,

$$vu_x - vu_x \cdot \left\{\frac{dx}{ds}\right\}^2 - vu_y \cdot \left\{\frac{dx}{ds} \cdot \frac{dy}{ds}\right\} - vu_z \cdot \left\{\frac{dx}{ds} \cdot \frac{dz}{ds}\right\} = \frac{d^2x}{ds^2}$$

$$vu_y - vu_x \cdot \left\{\frac{dy}{ds} \cdot \frac{dx}{ds}\right\} - vu_y \cdot \left\{\frac{dy}{ds}\right\}^2 - vu_z \cdot \left\{\frac{dy}{ds} \cdot \frac{dz}{ds}\right\} = \frac{d^2y}{ds^2}$$

$$vu_z - vu_x \cdot \left\{\frac{dz}{ds} \cdot \frac{dx}{ds}\right\} - vu_y \cdot \left\{\frac{dz}{ds} \cdot \frac{dy}{ds}\right\} - vu_z \cdot \left\{\frac{dz}{ds}\right\}^2 = \frac{d^2z}{ds^2}$$

Then,

$$V_x\left(\left(\frac{dx}{ds}\right)^2 - 1\right) + V_y \frac{dx}{ds} \cdot \frac{dy}{ds} + V_z \frac{dx}{ds} \cdot \frac{dz}{ds} = -\frac{d^2x}{ds^2}$$

$$V_x \frac{dy}{ds} \cdot \frac{dx}{ds} + V_y\left(\left(\frac{dy}{ds}\right)^2 - 1\right) + V_z \frac{dy}{ds} \cdot \frac{dz}{ds} = -\frac{d^2y}{ds^2}$$

$$V_x \frac{dy}{ds} \cdot \frac{dz}{ds} + V_y \frac{dz}{ds} \cdot \frac{dy}{ds} + V_z\left(\left(\frac{dz}{ds}\right)^2 - 1\right) = -\frac{d^2z}{ds^2}$$

where,

$$V_x = vu_x; \quad V_y = vu_y; \quad V_z = vu_z.$$

For the evaluation of V_x, V_y, V_z,

$$V_x = v \cdot \left(\frac{1}{v}\right)_x = \frac{1}{v} \cdot (-v_x) = -\left(\frac{v_x}{v}\right)$$

$$v = \left(\sqrt{x^2 + y^2 + z^2} - r_q\right) \cdot f_q + v_q; \text{ within } q^{th} \text{ band of radial velocity model}$$

$$v_x = f_q \cdot \frac{d\left(\sqrt{x^2 + y^2 + z^2}\right)}{dx}$$

$$V_x = -\left(\frac{f_q}{\left(\left(\sqrt{x^2 + y^2 + z^2} - r_q\right) \cdot f_q + v_q\right)}\right) \cdot \frac{x}{\sqrt{x^2 + y^2 + z^2}},$$

$$V_y = -\left(\frac{f_q}{\left(\left(\sqrt{x^2+y^2+z^2}-r_q\right)\cdot f_q+v_q\right)}\right)\cdot\frac{y}{\sqrt{x^2+y^2+z^2}},$$

$$V_z = -\left(\frac{f_q}{\left(\left(\sqrt{x^2+y^2+z^2}-r_q\right)\cdot f_q+v_q\right)}\right)\cdot\frac{z}{\sqrt{x^2+y^2+z^2}}.\ ]$$

Given this development, the ray-tracing algorithm can be provisionally formulated as follows:

1. The derivation of $\left(\dfrac{dx}{ds},\dfrac{dy}{ds},\dfrac{dz}{ds}\right)_{i=0}$ from the initial conditions

    (in other words, $\underline{r}_{i=0}$ and a hypothesised take-off angle $\psi_0|_{i=0}$)

2. The calculation of $V_{x,y,z}$ for this set of $\left(\dfrac{dx}{ds},\dfrac{dy}{ds},\dfrac{dz}{ds}\right)_i$ and $\underline{r}_i=(x,y,z)_i$

3. The forming $\left(\dfrac{d^2x}{ds^2},\dfrac{d^2y}{ds^2},\dfrac{d^2z}{ds^2}\right)$ from 2

4. By iteration:

$$\underline{r}_{i+1}=\underline{r}_i+h\cdot\left(\frac{dx}{ds},\frac{dy}{ds},\frac{dz}{ds}\right)_i+\frac{h^2}{2!}\cdot\left(\frac{d^2x}{ds^2},\frac{d^2y}{ds^2},\frac{d^2z}{ds^2}\right)_i\ ;$$

$$\left(\frac{dx}{ds},\frac{dy}{ds},\frac{dz}{ds}\right)_{i+1}=\left(\frac{dx}{ds},\frac{dy}{ds},\frac{dz}{ds}\right)_i+h\cdot\left(\frac{d^2x}{ds^2},\frac{d^2y}{ds^2},\frac{d^2z}{ds^2}\right)_i\ ;$$

5. Recycle to 2, using new $\underline{r}_{i+1}$ and $\left(\dfrac{dx}{ds},\dfrac{dy}{ds},\dfrac{dz}{ds}\right)_{i+1}$.

The point-to-point aspect is achieved by a binary chop search against the required latitude, with accuracies of hit of $\geq 1.0_{10}-05$ kilometres being recorded as great circle distances away from the target station.

All velocities required at all points in the developing trajectory of the ray are defined by a radial earth velocity model such as those given as ak135 and PREM models below.

The usual care is taken in interpreting the discontinuities and step functions present in this model. Discontinuities of the step kind are approached as

$$\lim_{d\to d_0^+}\left(v_0\right)=v_0^+$$

$$\lim_{d\to d_0^-}\left(v_0\right)=v_0^-$$

where d_0 is the abscissa value at the step discontinuity, and v_0^-, v_0^+ are the two values defining the step or jump in continuity.

It is worth mentioning that figures 5.32 to 5.37 give some credence to the veracity of the methods described here. They demonstrate a "P-wave shadow" between about 90 degrees of relative latitude and 143 degrees of relative latitude. This is corroborated in B. L. N. Kennett et al., which corroboration is also given in Figure 5.45 here.

Radial and Lagrangian Ray Trace Algorithm

The first algorithm of this pair is based on the existence of a ray constant in a radially symmetrical spherical space, expressed as

$$\frac{r_k \sin i_k}{v(r_k)} = \frac{r_{k+1} \sin i_{k+1}}{v(r_{k+1})} = \wp$$

This applies when the zone within which the pair r_k, r_{k+1} finds itself, is governed by a spatial velocity distribution with a linear law. In other words,

$$v(r) = mr + c$$

where

$$v(r) = v(r_k); \quad m = \frac{(V_1 - V_0)}{(R_1 - R_0)}; \quad r = r_k - R_0; \quad c = V_0$$

and the pairs (V_0, R_0) and (V_1, R_1) bound the zone (region) as closed or open.
Within the confines of a particular region with its concomitant law or spatial distribution

$$\sin i_{k+1} = \frac{v_{k+1}}{v_k} \cdot \frac{r_k}{r_{k+1}} \cdot \sin i_k$$

$\sin i_k$ represents the take-off angle at r_k in whatever region r_k resides.
The constructed ray path is driven by the radial change

$$r_{k+1} = r_k + d \cdot \Delta r$$

Here, d is a switching value $(+1.0, -1.0)$, depending on whether the ray is descending, (-1.0), or ascending, $(+1.0)$.

If the resulting value intended to define $\sin i_{k+1}$ is $> Unity$, then a search is initiated to attempt to establish a direction that will allow this sine value to lie within the limits $[-1.0, +1.0]$. If successful, the direction switch will be changed appropriately.

The angular displacement of the leading point of the ray is a function of the driving radius. The following development shows how

$$\hat{u} = \arcsin\left(\frac{r_k}{r_{k+1}} \cdot \sin i_k\right)$$

$$\hat{f} = \pi - \left(\hat{i}_k + \hat{u}\right)$$

Having found the incremental angular displacement, $\hat{f}$, for the kth leg. Then the corresponding ray path segment length is given by

$$\Delta s = \sqrt{r_k^2 + r_{k+1}^2 - 2 \cdot r_k r_{k+1} \cos \hat{f}}$$

The time for the ray to traverse this segment may be approximated by

$$\Delta t = \frac{\Delta s}{\left(\frac{\left(v(r_k) + v(r_{k+1})\right)}{2}\right)}$$

or by an integrative method that takes into account the radial constriction along the segment due to the ray inhabiting a spherical space. This method is described with the help of figure 5.1. The figure attempts to show, in a conceptual manner, that for a rectilinear segment crossing a radial (linearly changing with radius) velocity field, then the rate of change of the field for such a velocity model would not necessarily be uniform in the direction of the traversing segment.

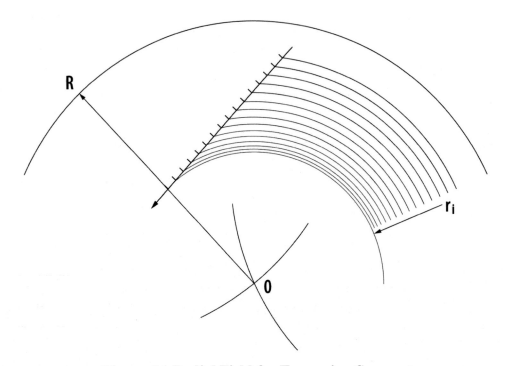

Figure 5.1 Radial Field for Traversing Segment

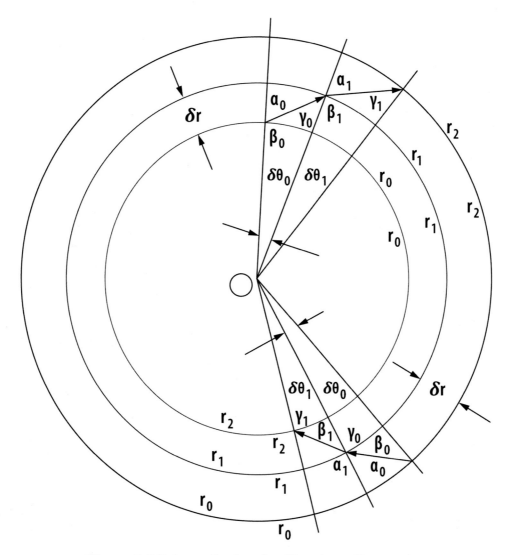

Figure 5.2 Schematic showing Ray-trace Parameters

The second method in this chapter, the Lagrange method, is such that $\delta\theta$ can be kept constant as the ray is progressed. Here (in figure 5.2) is described this main option for implementation of the algorithm. For the upward going ray, initially we have:

$$\beta_0 = \pi - \alpha_0;$$

$$\gamma_0 = \pi - \left(\beta_0 + \delta\theta_0\right)$$

$$r_1 = r_0 \cdot \frac{\sin\beta_0}{\sin\gamma_0}$$

$$\delta s_0 = r_1 \cdot \frac{\sin\delta\theta_0}{\sin\beta_0}$$

or

$$\delta s_0 = \sqrt{\left(r_0^2 + r_1^2 - 2r_0 r_1 \cdot \cos\left(\delta\theta_0 \right) \right)}$$

where

$$\delta\theta_0 \text{ is a constant}$$

then

$$\delta t_0 = \frac{2 \cdot \delta s_0}{\left(v\left(r_0 \right) + v\left(r_1 \right) \right)}$$

Initially, α_0 represents the supplied take-off angle. The δs_i are the line segments that collectively make up the ray.

The subsequent stage follows on with

$$\alpha_1 = \sin^{-1}\left(\sin\gamma_0 \cdot \left(\frac{v\left(r_1 \right)}{v\left(r_0 \right)} \right) \cdot \left(\frac{r_0}{r_1} \right) \right)$$

by refraction.

For downward going rays this process can be seen to have a natural reversal.

Point-to-Point Ray Tracing in Flat Strata with Lagrangian Control

It is hypothesised that arcs subtended on the earth's surface by angles of up to and including nine degrees can be thought of as delimiting a region of crustal strata, that could, for our purposes, be considered "flat".

The routines described in this Chapter are used to set up a tabular structure (to *some practicable depth*) across a set of "lateral distances" considered to lie on a flat surface. Such distances may be generated by

$$\lambda_i = 1.11195_{10}02 \cdot \beta_i$$

(assuming, for the sake of argument an earth radius $R_e = 6371$ km)

Here λ_i are the distances generated and β_i are angles in degree measure and $\beta_i \in \left[0^{\circ}, 4.5^{\circ} \right]$.. This range of angles is considered relative to an epicentre at *Zero* degrees.

Such a tabular structure could be instrumental in the very rapid determination of a "shallow" hypocentre by close-lying, proximal stations (in other words, within 4.5 $^{\circ}$ of the epicentre), using an interpolative scan technique identical to the one outlined in the main text.

The control of the P2P table-generating algorithm can be effected by using two interlocking multiplying systems. These systems sequentially modify the initial take-off angle, ψ, such that the ray in question, starting in the lower reaches of the flat strata, will eventually arrive at the surface at any requested lateral distance λ_i within the prescribed interval above.

The system where two interlocking multiplying controls are implemented is here called the *height/length method*. However, there is an alternative system that only closes the feedback loop relative to lateral distance, or length. This is referred to as the *length method* and is described next.

The multiplying systems involve feedback in the form of the current divergence from the present target:

$$\lambda = l + \varepsilon_\lambda \cdot l = l \cdot \left(1 + \varepsilon_\lambda\right)$$

Here, λ is the intended target distance, and l is the current value achieved by the ray for a given value of ψ. We write

$$\varepsilon_\lambda\big|_i = \frac{\lambda - l}{l}$$

$$\psi_{i+1} = \psi_i \cdot \left(1 - \frac{\varepsilon_\lambda\big|_i}{n_i}\right)$$

where $\varepsilon_\lambda\big|_i$ is the multiplying factor at the juncture between the ith and the $\left(i+1\right)^{th}$ iteration. n_i is the number of steps the ray tracer portion of the algorithm has taken to achieve the off-target position l at the ith iteration.

The height/length method is predicated on whether the surface is reached first or whether the required length is reached before the surface. If the surface is reached before the target length, then we apply

$$\lambda = l + \varepsilon_\lambda \cdot l = l \cdot \left(1 + \varepsilon_\lambda\right)$$

$$\varepsilon_\lambda\big|_i = \frac{\lambda - l}{l}$$

$$\psi_{i+1} = \psi_i \cdot \left(1 \pm \frac{\varepsilon_\lambda\big|_i}{n_i}\right)$$

Here, the sign of $\varepsilon_\lambda\big|_i$ is dependent on whether ψ should contract or expand.

In this case, $\varepsilon_Z\big|_i = Zero$, since the surface has been exactly reached or surpassed. Then if the length is reached before the surface

$$Z = z \cdot \left(1 + \varepsilon_Z\right)$$

$$\varepsilon_Z\big|_i = \frac{Z - z}{z}$$

$$\psi_{i+1} = \psi_i \cdot \left(1 \pm \frac{\varepsilon_Z\big|_i}{n_i}\right)$$

The take-off angle, ψ, must be reduced to give an upward going ray. In this case, $\varepsilon_\lambda\big|_i = Zero$, since the length has been equalled or surpassed.

By observation, both $|\varepsilon_\lambda|$ and $|\varepsilon_z|$ tend to *Zero* in the limit, as is exemplified by Figures 5.3 and 5.4.

In general, $|\varepsilon_\lambda|$ and $|\varepsilon_z|$ converge to *Zero* most rapidly in the height/length method, while this convergence of $|\varepsilon_\lambda|$ alone in the length method is slower.

Next, we'll discuss the main programs that are vehicles for these two kinds of Lagrangian control. First up is

"TimeBased01.cpp"

This program uses routines of both kinds to lay down the four part data tables used in the tabular interpolation scans over a crustal set of strata considered flat.

Another such program is

"TimeBased04.cpp"

This program uses, again, routines of both kinds to perform self-tests against the data structures created by "TimeBased01.cpp".

Also, mention can be made of "TRTST 00.cpp" and "TRTST 01.cpp", which furnish individual ray paths for given initial conditions. They are part of an exercise to try to transform ray paths lying in a flat rectilinear distribution of strata to a stratification distributed radially within an annulus.

The Lagrangian routines for use in flat stratification are contained in the file

"HypocentreScanningTransBased.h"

Each of the two methods is worked as a two-tier structure with a controlling layer calling a ray tracer. Each controlling layer performs the multiplicative control on the take-off angles, ψ, which are passed to the ray tracers. The ray tracers themselves are open loop, while the multiplicative controllers close this loop. The control level is activated by a zeroth, or calling level, formed by the main programs mentioned above, and passes back:

- Path time for rays
- Path length of rays
- Final take-off angles
- Target lateral distances
- Accuracy measures (in other words, the errors in reaching the required lateral distances)

This information is passed back for each case, which has in turn been passed back from the last call on the ray tracers.

The tracer routines concerned are, by name

```
void LagrangePointToPointX01(…)
void LagrangePointToPointZ01(…)
```

These are controlling levels for the length method and the length/height method respectively. They are called from "TimeBased01.cpp" and generate output of data structures (see above). The second set of controlling routines is

void LagrangePointToPoint01X(…)
void LagrangePointToPoint01Z(…)

These are similarly called from the self-test program TimeBased04.cpp and generate no output. Both pairs of controllers call the ray tracers

void RayPathQDXX(…)
void RayPathQDXZ(…)

which correspond to the two methods—control by length and by length/height respectively. These ray tracers are parameterised by the initial take-off angle, ψ. And so controlled in this context, they are identical to the tracers described under the label "Lagrangian" above.

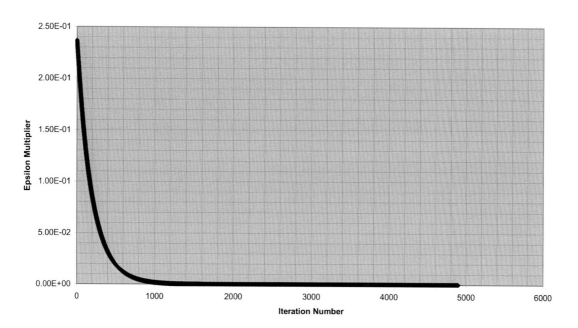

Figure 5.3

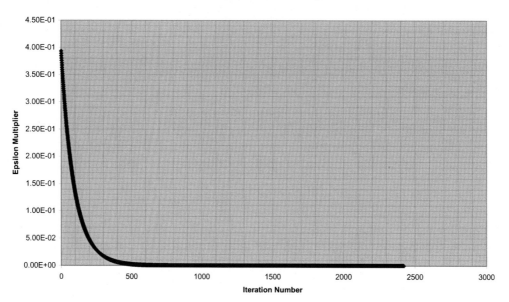

Figure 5.4a

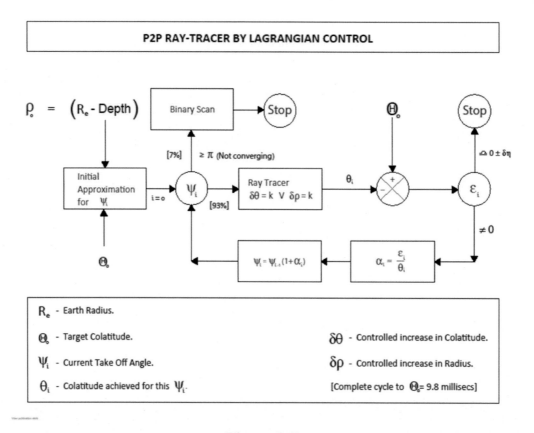

Figure 5.4b

The "Lagrangian" system pictured as a P2P Ray-tracer at **Figure 5.4b**, above, is for use on a Spherical Earth model and ir is found to be an extremely fast and accurate method for producing the Tables to be used for any form of tabular scan (see Chapter 6 et seq.)

These Tables, as referred to above, consist of:

• Travel -times
• Take-off Angles
• Accuracy Levels
• Path lengths
• Error Log.

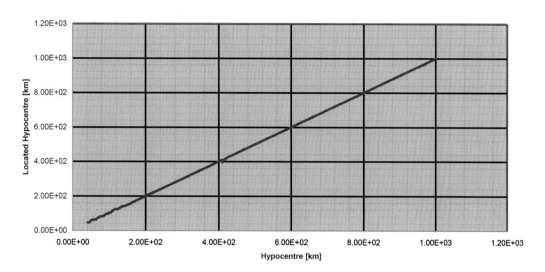

Figure 5.5

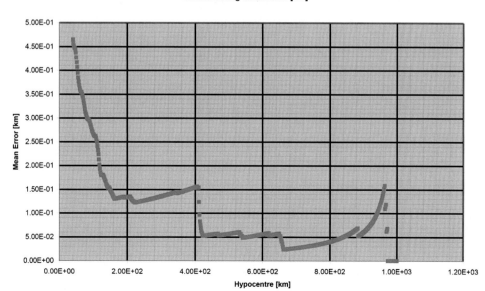

Figure 5.6

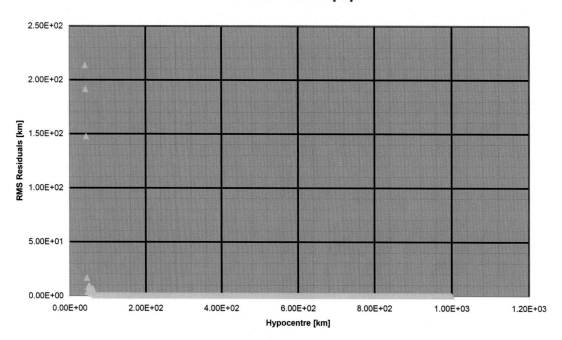

Figure 5.7

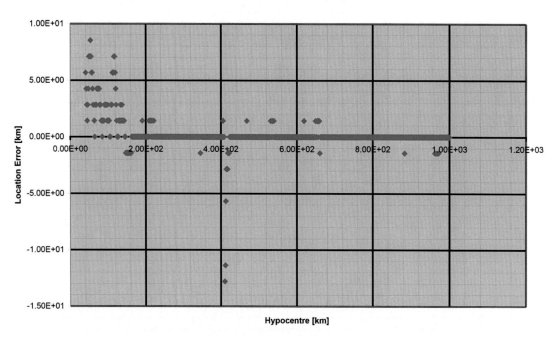

Figure 5.8

Figures 5.5 to 5.6 are the results of a self-test based on tables of ray path times and take-off angles generated by "TimeBased01.cpp" as has been described above. The self-test was run by "TimeBased04.cpp".

The tables were constructed for a hypothetical and parallel-sided region of 5 to 1,000 kilometres in depth and a region of $0.1°$ to $9°$ in lateral distance, using the methods mentioned above and in the main text.

The method currently collapses at an upper limit of around 40 kilometres depth, owing to inadequate precision and sparse amounts of data relating to the fine structure of strata at these levels.

Consider again the recurrence relation:

$$\psi_{i+1} = \psi_i \left(1 + \frac{\varepsilon}{n} \right)$$

In this case, we relate to the previous definition of ε:

$$\varepsilon_\lambda = \frac{\lambda - l}{l} = \frac{\lambda - K(\psi_i)}{K(\psi_i)}$$

Function definitions are taken to be as follows:

$$K(\psi_i) = \delta z_0 \tan\left(\psi_i|_0\right) + \sum_{j=1}^{n-1} \delta z_j \tan\left(\psi_i|_j\right)$$

$$\frac{dK(\psi_i)}{d\psi_i}(\psi_i) = \delta z_0 \sec^2\left(\psi_i|_0\right) + \sum_{j=1}^{n-1} \delta z_j \sec^2\left(\psi_i|_j\right) \cdot \left(\frac{dT}{d\psi_i}\right)_j$$

These are the naive ray tracer, K, and its derivative, $\dfrac{dK}{d\psi_i}$. They are supported by

$$\left(\frac{dT}{d\psi_i}\right)_j = \cos\left(\psi_i|_0\right) \cdot \prod_{k=0}^{j-1} \frac{1}{\sqrt{\left(\dfrac{v_k}{v_{k+1}}\right)^2 - \sin^2\left(\psi_i|_k\right)}} = \cos\left(\psi_i|_k\right)$$

and

$$\psi_i|_k = \sin^{-1}\left(\sin\left(\psi_i|_{k-1}\right) \cdot \left(\frac{v_k}{v_{k-1}}\right) \right)$$

$$\psi_i|_1 = \sin^{-1}\left(\sin\left(\psi_i|_0\right) \cdot \left(\frac{v_1}{v_0}\right) \right)$$

Here, $\left(\delta z_i, v_i\right)$ is an incremental layer depth and velocity pair. $\psi_i|_0$ is i^{th} incremental initial take-off angle for any given ray. Also the dot convention is to represent differentiation with respect to $\psi_i|_0$.

Now, let's look at this substituting

$$\psi_{i+1} = \frac{\lambda}{n \cdot K(\psi_i)} \cdot \psi_i + \left(\frac{n-1}{n}\right) \cdot \psi_i$$

An example is given at Figure 5.10. When and if

$$\lim_{i \to \infty} K(\psi_i) = \lambda$$

then the above expression for the recurrence becomes

$$\psi_{i+1} = \frac{\psi_i}{n} + \left(\frac{n-1}{n}\right) \cdot \psi_i$$

This is saying $\psi_{i+1} = \psi_i = \Psi$ in perpetuity, and we have a convergence of this recurrence relation, iff $K(\Psi) = \lambda..$

Further,

$$\psi_{i+1} = \psi_i \left(Unity - \frac{\lambda - K(\psi_i)}{K(\psi_i)} \right)$$

$$Unity - \frac{\psi_{i+1}}{\psi_i} = \frac{\lambda - K(\psi_i)}{K(\psi_i)}$$

Hence,

$$\lambda = K(\psi_i) + K(\psi_i) \cdot \left(Unity - \frac{\psi_{i+1}}{\psi_i} \right)$$

and so as $\psi_i \to \psi_{i=1} \to \Psi$, then:

$$\lambda = K(\psi_i) = K(\Psi)$$

So we can state that there is an equivalence relation:

$$K(\psi_i) = \lambda \iff \lim_{i \to \text{very large}} \{\psi_i\} = \Psi$$

If the series ψ_i converges, then $K(\psi_i)$ converges to λ.
We can write the above recurrence relation as

$$\psi_{i+1} = f(\psi_i)$$

We know that, if ψ_i takes a value equal to ψ^* and $\psi^* \in I$, an interval of convergence on the real line and $\left| \dfrac{df(\psi^*)}{d\psi_i} \right| < Unity$, then the recurrence relation will converge to some value, Ψ, for any initial value of ψ_i taken from I.

Here, as stated, the dot convention implies differentiation with respect to ψ_i.

$$f(\psi_i) = \frac{\lambda}{n \cdot K(\psi_i)} \cdot \psi_i + \left(\frac{n-1}{n}\right) \cdot \psi_i$$

$$\frac{df(\psi_i)}{d\psi_i} = \frac{\lambda}{nK(\psi_i)} \left(Unity - \psi_i \cdot \frac{\dfrac{dK(\psi_i)}{d\psi_i}}{K(\psi_i)} \right) + \left(\frac{n-1}{n}\right)$$

An example is given in Figures 5.11 and 5.12.

We also know that the rate of convergence is directly proportional to the value of $\dfrac{df(\psi_i)}{d\psi_i}$, where $\psi_i \to \Psi$.

We write

$$\varepsilon_{i+1} = \frac{df(\psi_i)}{d\psi_i} \cdot \varepsilon_i$$

where

$$\varepsilon_{i+1} = (\Psi - \psi_{i+1}) \wedge \varepsilon_i = (\Psi - \psi_i)$$

Now we have the value of Ψ 56 degrees, either from the equation $K(\psi_i) - \lambda = Zero$ or read off, as in this case, from Figure 5.13. Now instating Ψ into $\dfrac{df(\psi_i)}{d\psi_i}$, we get

$$\frac{df(\Psi)}{d\psi_i} = 0.866$$

which is a limiting value of Psi for the differential to be *Zero*.

Thus, ψ_i *will* converge, indeed over a broad range. So the function $K(\psi_i)$ will also go to λ, and our equivalence statement, suggested above, holds.

Taking the initial expression for $\dfrac{df(\psi_i)}{d\psi_i}$ above:

$$\left| \frac{\lambda}{nK(\psi_i)} \left(1 - \psi_i \cdot \frac{\dfrac{dK(\psi_i)}{d\psi_i}}{K(\psi_i)} \right) + \left(\frac{n-1}{n} \right) \right| < Unity$$

$$\left| \frac{\lambda}{K(\psi_i)} \left(1 - \psi_i \cdot \frac{\dfrac{dK(\psi_i)}{d\psi_i}}{K(\psi_i)} \right) \right| < Unity$$

Assume $\psi_i \to \Psi$, then we have:

$$\left| 1 - \frac{\Psi}{\lambda} \cdot \frac{dK(\Psi)}{d\psi_i} \right| < Unity$$

For a graph of this function, see Figure 5.13 Note the zero crossing at $56\,^{\circ}$.
Note it's at the limit

$$\frac{\lambda}{\Psi} \geq \frac{dK(\Psi)}{d\psi_i} > Zero$$

Since we do not know, a priori, the value of Ψ, we could investigate the above inequality to establish the interval within which Ψ will lie—in other words, all ψ_i, such that

$$\left(\frac{\lambda}{\psi_i} \right) > \left| \frac{\lambda}{\psi_i} - \frac{dK(\psi_i)}{d\psi_i} \right| \geq Zero].$$

Given the above development it is possible to tabulate the convergence properties within the intervals

$$\lambda \in \left[0.1\,^{\circ},\ 9.0\,^{\circ} \right]$$

$$Depth \in \left[40 \ (say),\ 1000 \right] km$$

for any given or known crustal stratigraphic structure, to use the means of finding the hypocentral depth proposed in this.

In the case of all of the following graphs, the problem they demonstrate is parameterised by a lateral distance (λ) of four degrees and a depth of 400 kilometres.

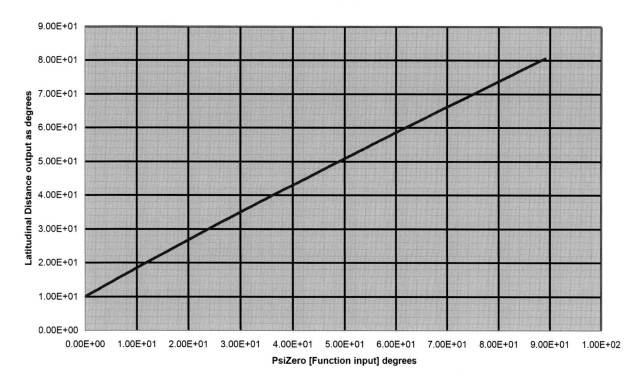

Figure 5.9

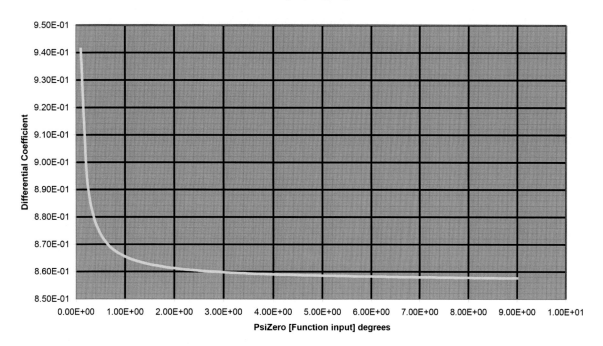

Figure 5.10

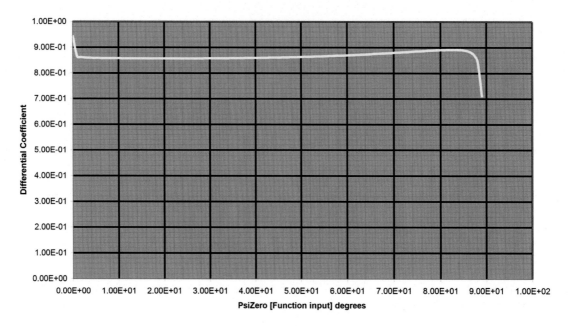

Figure 5.11

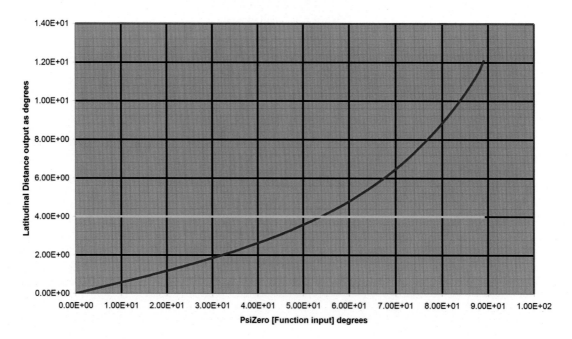

Figure 5.12

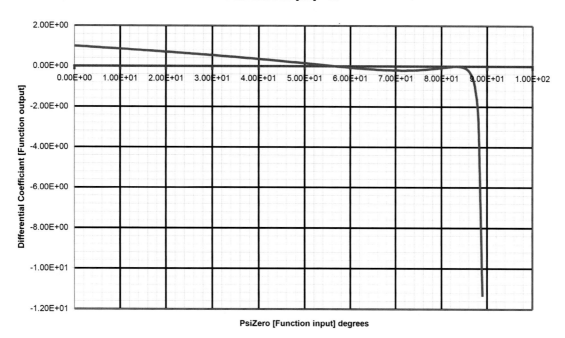

Figure 5.13

Database of Tabular Structures

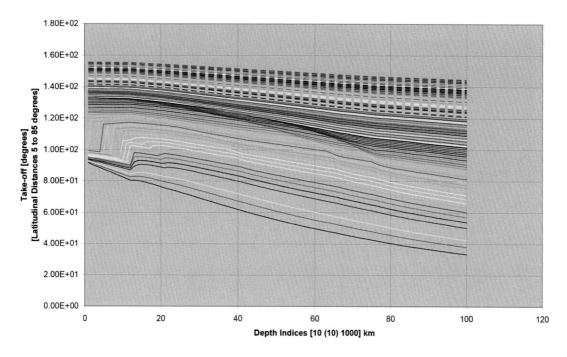

Figure 5.14

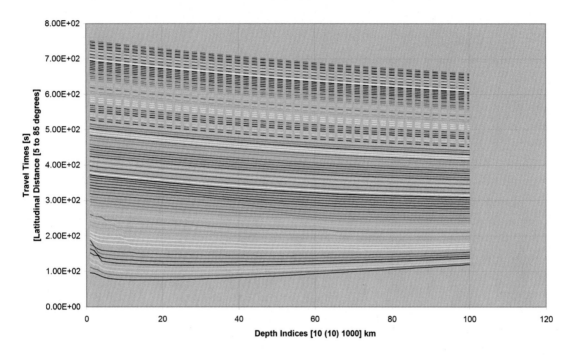

Figure 5.15

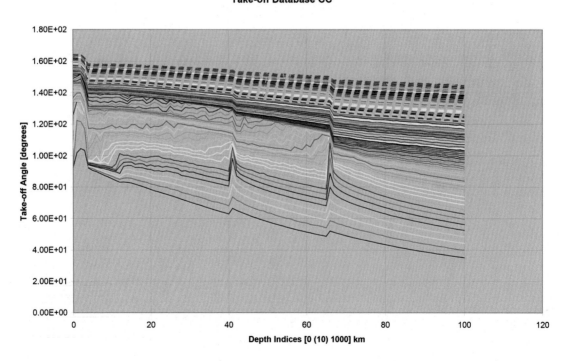

Figure 5.16

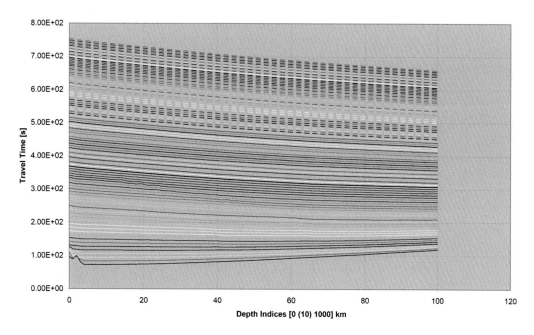

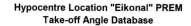

Figure 5.17

Figures 5.15 and 5.16 represent the smoothed output from "ETTST 04.cpp" working on travel times and take-off angles produced by the eikonal ray tracer.

Likewise, figures 5.17 and 5.18 are the output from "ETTST 04.cpp" working on the data produced by the radial ray tracer.

All these cases use the earth radial velocity model, ak135.

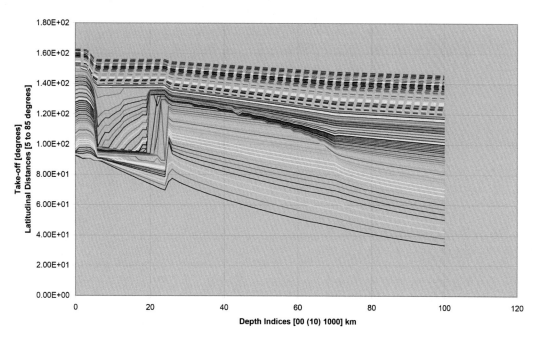

Figure 5.18

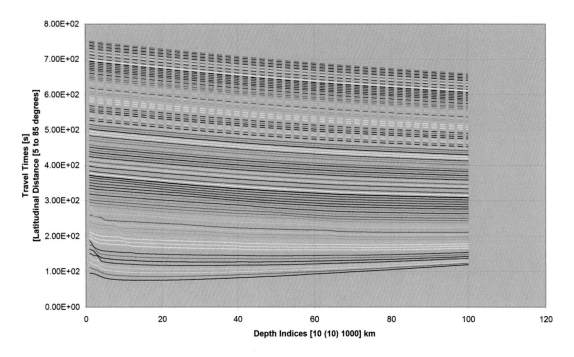

Figure 5.19

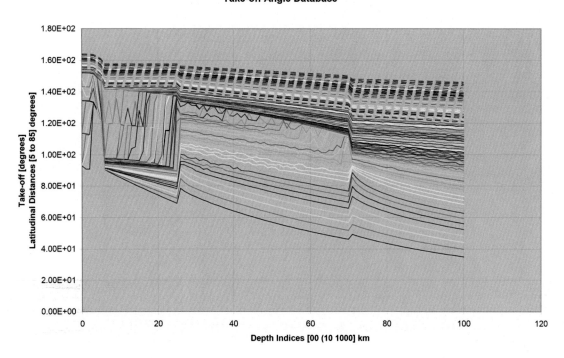

Figure 5.20

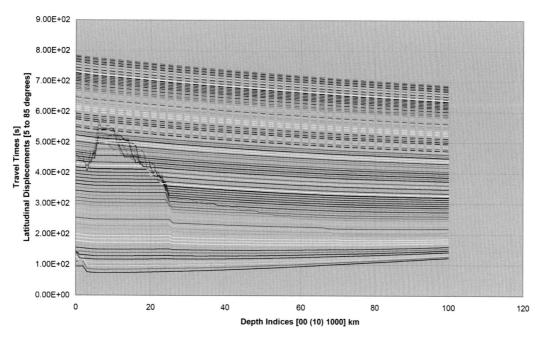

Figure 5.21

Figures 5.19 and 5.20 represent the smoothed output from "ETTST 04.cpp" working on travel times and take-off angles produced by the eikonal ray tracer.

Likewise, figures 5.21 and 5.22 are the output from "ETTST 04.cpp" working on the data produced by the radial ray tracer.

All these latter cases use the earth radial velocity model "PREM" using Lagrange interpolation.

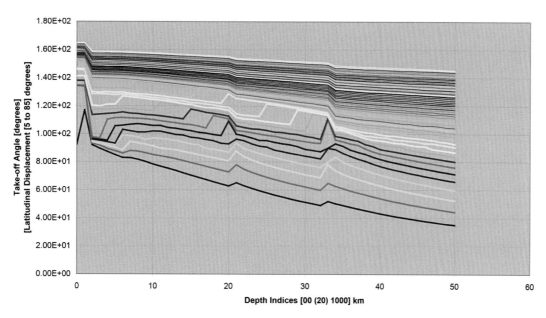

Figure 5.22

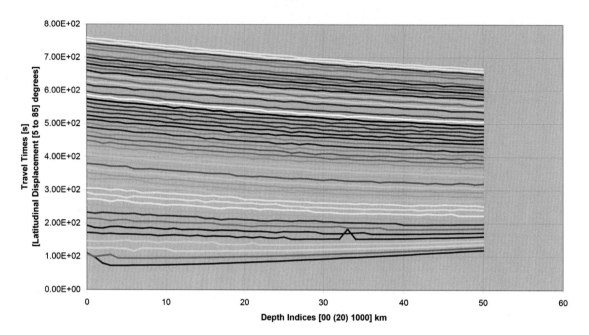

Hypocentre Location "Radial" [ak135]
Travel Time Database
[51 by 51]
UnInterpolated

Figure 5.23

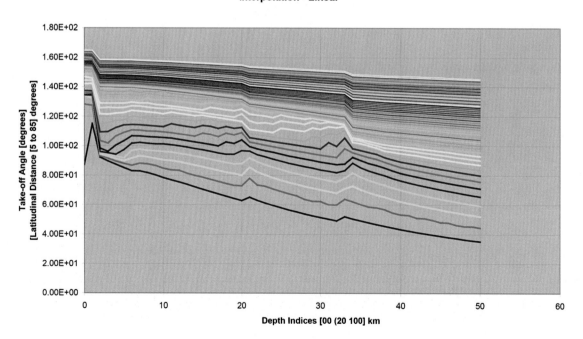

Hypocentre Location "Radial" [ak135]
Take-off Angle Database
[51 by 51]
Interpolation - Linear

Figure 5.24

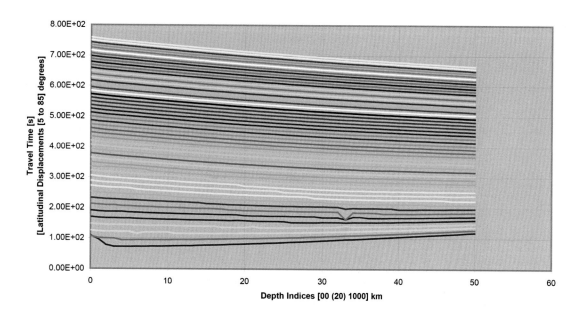

Figure 5.25

Figures 5.22 and 5.23 represent some unsmoothed output consisting of travel times and take-off angles produced by the radial ray tracer.

Figures 5.24 and 5.25 are the output from "ETTST 04 (05).cpp" working on the data produced by the radial ray tracer.

All these latter cases use the earth radial velocity model, ak135 .

The smoothed output referred to in the above, from "ETTST 04 (05).cpp", was in the form of linear interpolation.

Ray Traces

In the following Figures (5.26 to 5.44) the Independent Variable axis represents the radial distance from Earth center, while the Abscissa axis represents the polar angle subtended fron Earth center.

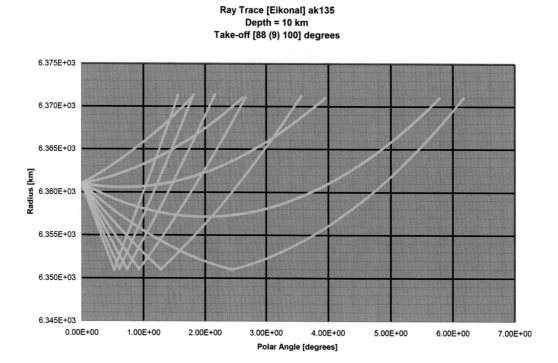

Figure 5.26

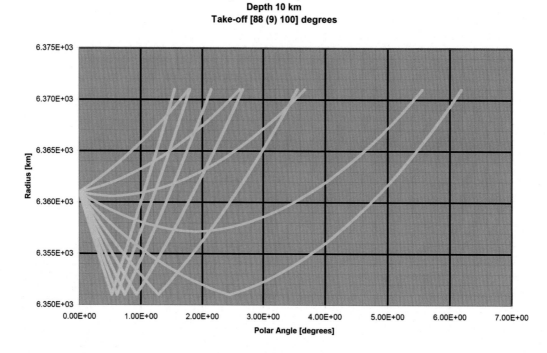

Figure 5.27

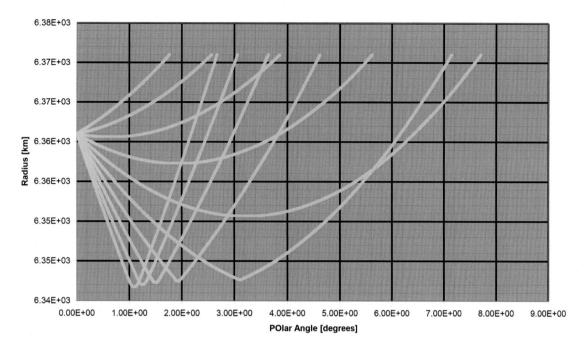

Figure 5.28

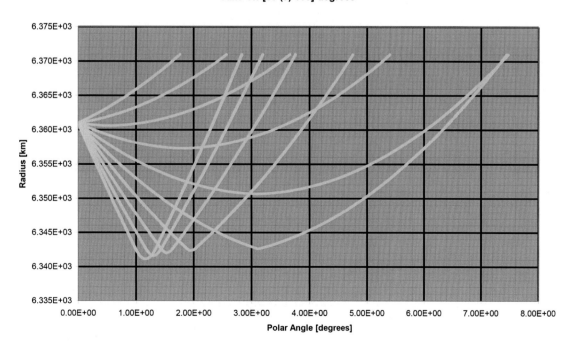

Figure 5.29

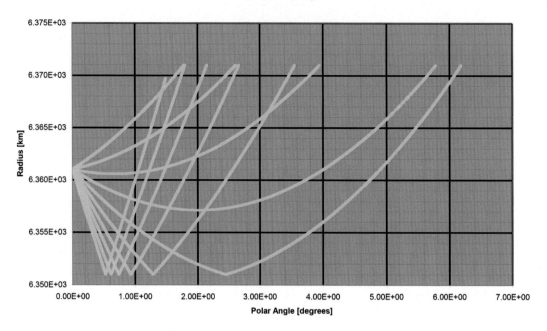

Figure 5.30

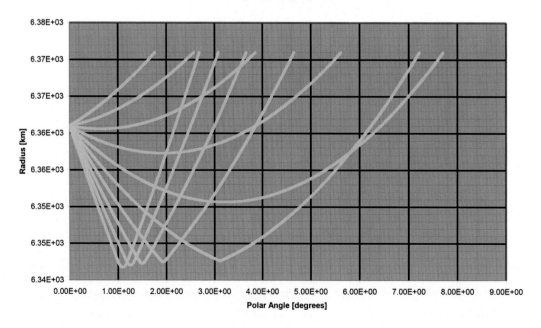

Figure 5.31

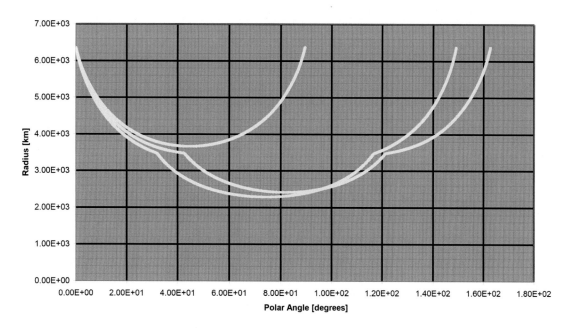

Figure 5.32

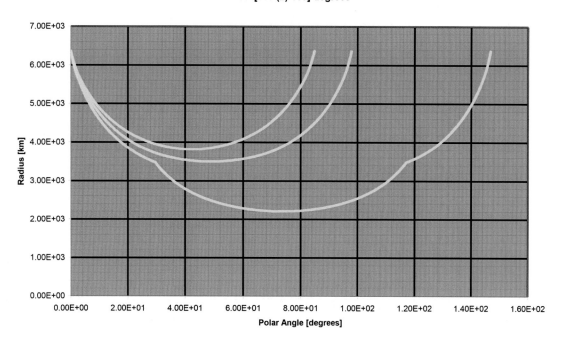

Figure 5.33

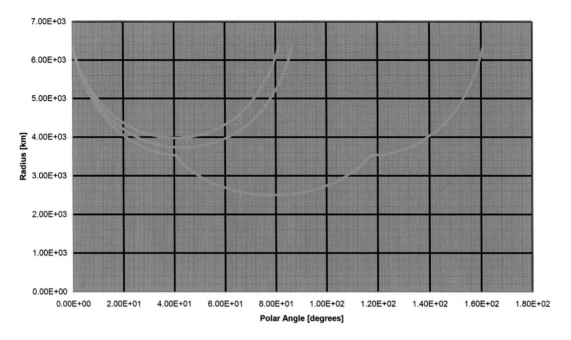

Figure 5.34

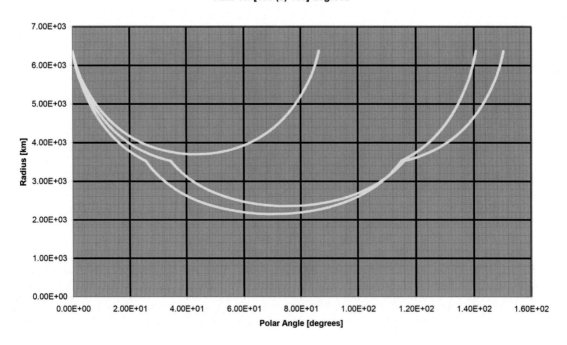

Figure 5.35

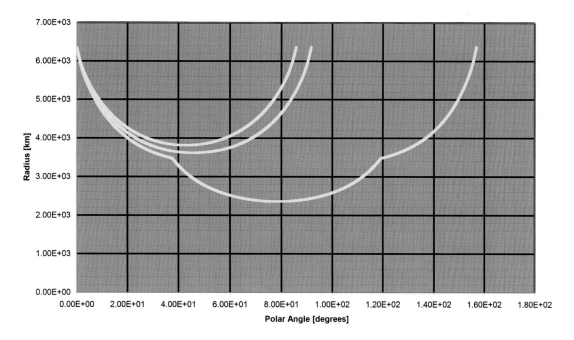

Figure 5.36

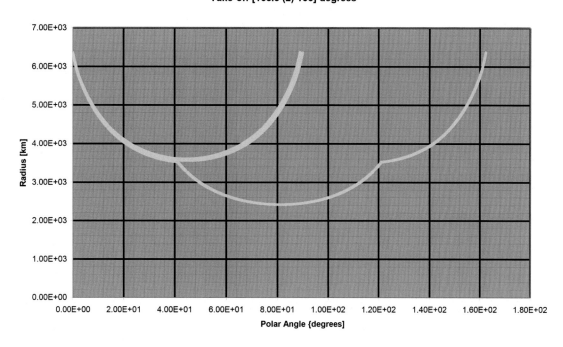

Figure 5.37

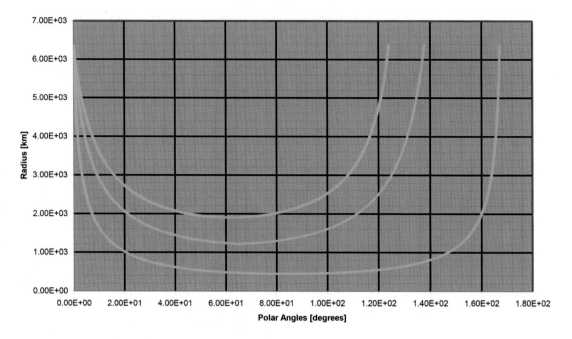

Figure 5.38

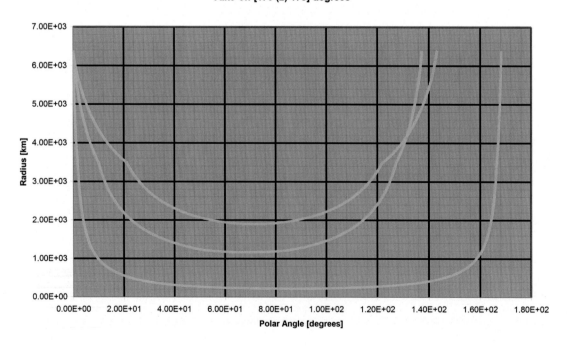

Figure 5.39

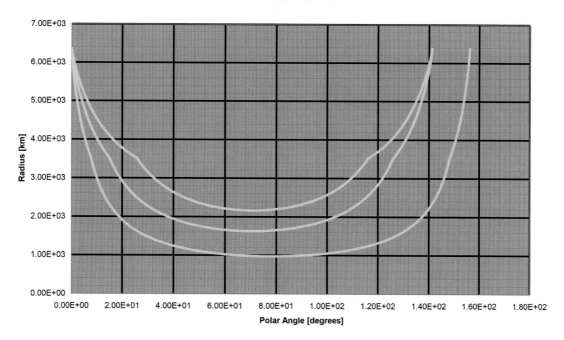

Figure 5.40

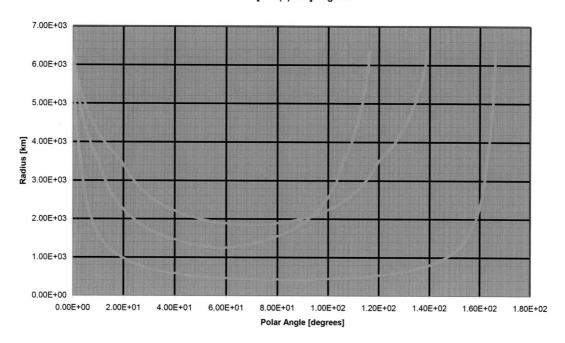

Figure 5.41

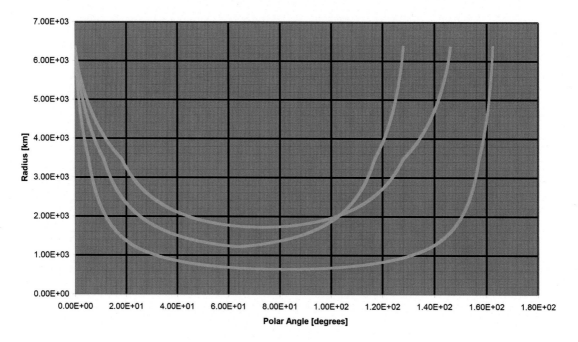

Figure 5.42

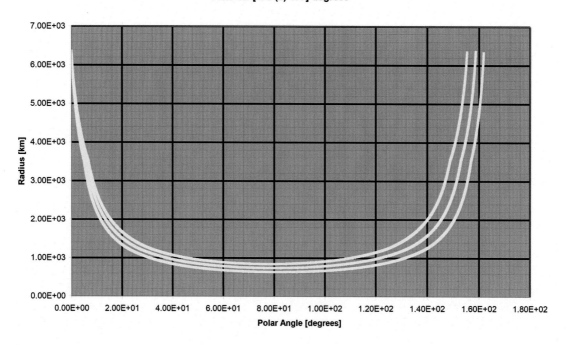

Figure 5.43

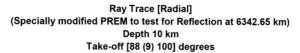

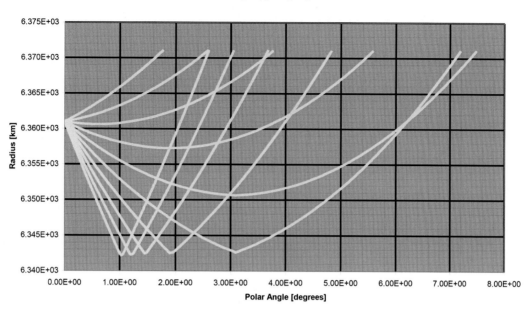

Figure 5.44

In this series of Figures, (5.26 to 5.44), the velocity models used were the PREM[11] and ak135[12] The convention for denumerating the sequences of take-off angles used was:

(Lowest angle [number of gaps] greatest angle)

All angles are given in degrees and all lengths in kilometres. Figure 5.45 represents the overall template into which the above three groups of ray traces should fit. The first group (figures 5.26 to 5.31) represent reflections in the "near field". The second group (figures 5.32 to 5.37) represents the inception of the P-wave "shadow". The third (figures 5.38 to 5.43) represents the "far field" propagation. Figure 5.44 is to be taken by itself, and represents a "beefing up" of the reduced PREM model to supply a reflecting horizon (at radius 6342.65 km) for the source depth of 10 km.

The ak135 Earth velocity model, together with a reduced version of the PREM (preliminary reference earth model) follows:

```
135

5.8        3.460      2.720       6371.0
5.8        3.460      2.720       6351.0
6.5        3.850      2.920       6351.0
6.5        3.850      2.920       6336.0
8.040      4.480      3.320       6336.0
8.045      4.490      3.345       6293.5
```

---

[11] Anderson & Djiewonsky, 1980

[12] Kennett et al. 1995.

8.050	4.500	3.371	6251.0
8.175	4.509	3.398	6206.0
8.300	4.518	3.426	6161.0
8.300	4.523	3.426	6161.0
8.483	4.609	3.456	6111.0
8.665	4.696	3.486	6061.0
8.847	4.783	3.517	6011.0
9.030	4.870	3.547	5961.0
9.360	5.080	3.756	5961.0
9.528	5.186	3.818	5911.0
9.696	5.292	3.879	5861.0
9.864	5.398	3.941	5811.0
10.032	5.504	4.003	5761.0
10.20	5.610	4.065	5711.0
10.79	5.960	4.371	5711.0
10.923	6.090	4.401	5661.0
11.056	6.209	4.431	5611.0
11.135	6.243	4.489	5561.5
11.222	6.280	4.486	5512.0
11.307	6.316	4.517	5462.5
11.391	6.351	4.546	5413.0
11.470	6.385	4.574	5363.5
11.550	6.419	4.603	5314.0
11.627	6.451	4.631	5264.5
11.703	6.483	4.659	5215.0
11.777	6.514	4.687	5165.5
11.849	6.544	4.715	5116.0
11.920	6.573	4.724	5066.5
11.990	6.601	4.770	5017.0
12.058	6.628	4.797	4967.5
12.125	6.655	4.825	4918.0
12.191	6.681	4.852	4868.5
12.255	6.707	4.878	4819.0
12.318	6.733	4.905	4769.5
12.382	6.757	4.932	4720.0
12.443	6.781	4.958	4670.5
12.503	6.805	4.985	4621.0
12.563	6.829	5.011	4571.5
12.622	6.852	5.037	4522.0
12.680	6.874	5.063	4472.5
12.738	6.897	5.089	4423.0
12.796	6.919	5.114	4373.5
12.853	6.942	5.140	4324.0
12.910	6.963	5.165	4274.5
12.967	6.985	5.190	4225.0

13.022	7.006	5.215	4175.5
13.078	7.028	5.240	4126.0
13.134	7.050	5.265	4076.5
13.189	7.072	5.290	4027.0
13.247	7.093	5.314	3977.5
13.302	7.114	5.339	3928.0
13.358	7.137	5.363	3878.5
13.416	7.159	5.387	3829.0
13.474	7.181	5.411	3779.5
13.531	7.203	5.435	3731.0
13.590	7.226	5.458	3681.0
13.649	7.249	5.482	3631.0
13.653	7.260	5.505	3581.3
13.657	7.270	5.528	3531.7
13.660	7.281	5.551	3479.5
8.000	0.000	9.915	3479.5
8.038	0.000	9.994	3431.7
8.128	0.000	10.072	3381.3
8.221	0.000	10.149	3331.0
8.312	0.000	10.223	3280.7
8.400	0.000	10.296	3230.3
8.486	0.000	10.368	3180.0
8.569	0.000	10.438	3129.7
8.650	0.000	10.506	3079.4
8.728	0.000	10.573	3029.0
8.804	0.000	10.639	2978.7
8.876	0.000	10.702	2928.4
8.946	0.000	10.765	2878.0
9.014	0.000	10.826	2827.7
9.079	0.000	10.885	2777.4
9.143	0.000	10.943	2727.0
9.204	0.000	11.000	2676.7
9.263	0.000	11.056	2626.4
9.321	0.000	11.109	2576.0
9.376	0.000	11.162	2525.7
9.430	0.000	11.214	2475.4
9.481	0.000	11.264	2425.1
9.531	0.000	11.313	2374.7
9.578	0.000	11.360	2324.4
9.623	0.000	11.407	2274.0
9.667	0.000	11.452	2223.7
9.710	0.000	11.496	2173.4
9.751	0.000	11.539	2123.1

9.791	0.000	11.581	2072.7
9.830	0.000	11.622	2022.4
9.868	0.000	11.661	1972.1
9.905	0.000	11.700	1921.7
9.941	0.000	11.737	1871.4
9.976	0.000	11.774	1821.1
10.010	0.000	11.809	1770.7
10.044	0.000	11.844	1720.4
10.077	0.000	11.877	1670.1
10.142	0.000	11.941	1569.4
10.174	0.000	11.972	1519.1
10.205	0.000	12.000	1468.8
10.233	0.000	12.031	1418.4
10.247	0.000	12.059	1368.1
10.274	0.000	12.087	1317.8
10.285	0.000	12.113	1267.4
10.289	0.000	12.139	1217.5
11.043	3.504	12.704	1217.5
11.059	3.519	12.729	1166.4
11.072	3.531	12.753	1115.7
11.085	3.543	12.776	1065.0
11.098	3.555	11.098	1014.2
11.117	3.566	12.819	963.5
11.132	3.576	12.839	912.8
11.146	3.586	12.857	862.1
11.159	3.596	12.875	811.4
11.172	3.604	12.892	760.7
11.183	3.613	12.907	710.0
11.194	3.620	12.922	659.3
11.213	3.634	12.947	557.8
11.222	3.640	11.222	507.1
11.229	3.645	11.229	456.4
11.236	3.650	11.236	405.7
11.242	3.654	11.242	355.0
11.248	3.658	12.993	304.3
11.252	3.661	12.999	253.6
11.256	3.663	13.004	202.9
11.259	3.665	13.007	152.1
11.261	3.667	13.010	101.4
11.262	3.668	13.012	50.7
11.262	3.668	13.012	0.0

Earth velocity model ak135: "alpha" (P-wave velocity), "beta" (S-wave velocity), and "rho" (density), as given, against earth radius by table 3 in "Constraints on Seismic Velocities in the Earth from Traveltimes", BLN Kennet et al., Geophysical Journal International, Vol. 122 issue 1, Blackwell 1995.

A reduced form of the PREM earth velocity model also follows:

```
17

5.833           6367.65
5.834           6342.65
7.00            6317.65         SURFACE
8.05            6317.65
7.90            6117.65


SURFACE REGION BOUNDARY
8.35            6117.65
9.10            5917.65         UPPER MANTLE
10.00           5667.65


MANTLE TRANSITION ZONE
10.60           5667.65         LOWER MANTLE
13.30           3527.65


CORE-MANTLE BOUNDARY
8.00            3517.65
9.50            2117.65         OUTER CORE
10.40           1242.65


INNER-OUTER CORE BOUNDARY
11.00           1242.65
11.15           617.65          INNER CORE
11.175          367.65
11.2            0.0

                                EARTH CENTRE
```

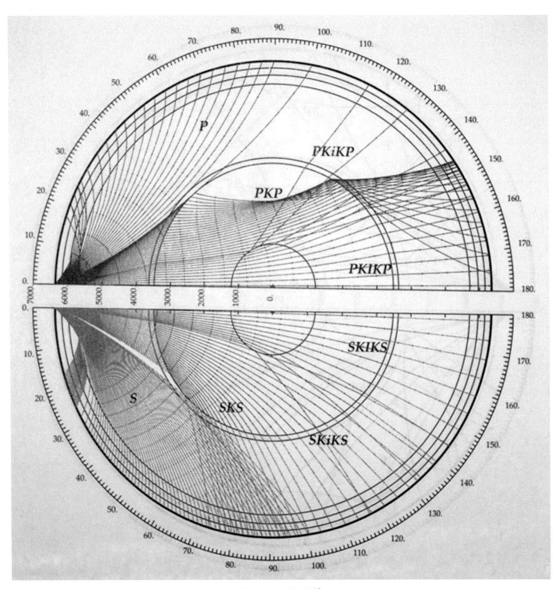

Figure 5.45[13]

[13]   This Figure is taken, with permission, from: BLN Kennett, *The Seismic Wavefield* (Figure 5.2, p80), Cambridge, 2001.

Chapter 6

Rapid and Concurrent Localisation of Epicentre and Hypocentre Pairs

The two kinds of interpolative tabular scans are described in this chapter. The first provides a hypocentre location given the epicentral coordinates of the event. The second represents a simultaneous, or concurrent, location of an epicentre / hypocentral pair using P-wave first arrivals and without prior knowledge of the epicentre. Each of these methods can, in fact, use P- or S-wave first arrivals. And each method uses a set of first arrival tables constructed by point-to-point (P2P) ray tracers, following any of a given set of earth velocity models. For graphical representations of such tables see figures 6.1 and 6.2.

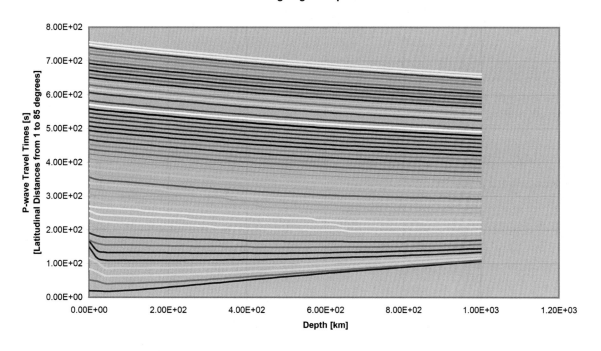

Figure 6.1 Travel time (first arrivals) database

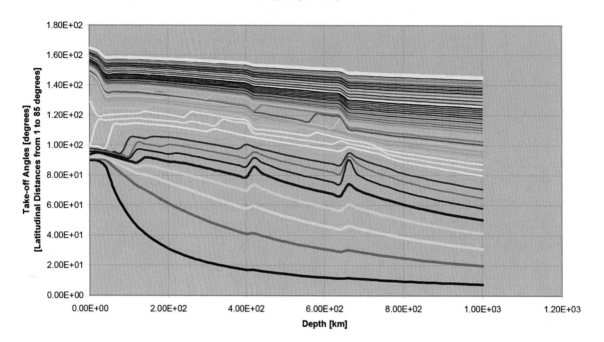

**Hypocentre Location "Lagrange" [Delta Theta = Constant]
[ak 135]
Take-off Angle Database [51 by 51]
P-wave: Lagrange Interpolation**

Figure 6.2 Take-off angle database

It is found for the implementation of each method that, with regard to speed, the first type (*epicentre known*) can perform a single location using fifteen to twenty stations, within 0.25 seconds, scanning 2,000 depth points from a prescribed depth. Normally, four scans are performed in tandem, exercising combinations of linear, cubic, and Lagrangian interpolation methods. For a track of such a scan, see figure 6.3. The second type (*with epicentre unknown*) will find both the epicentre and hypocentre of an event, using up to twenty stations, within 4 seconds. This scheme uses only linear interpolation and also produces clear minima (figure 6.3). However, the reason for the relative slowness of the latter routines is the larger amount of time spent in evaluating a "cost function" ("error indicator"), which monitors the fit (internal consistency) of an epicentre calculation, together with the match on a hypocentral depth. These timings are based on the throughput for a 3.2 megahertz machine.

Both methods can rapidly find results in close agreement with currently received values and, owing to their speed, might form strong assets as components in the context of an earthquake early warning (EEW) scheme.

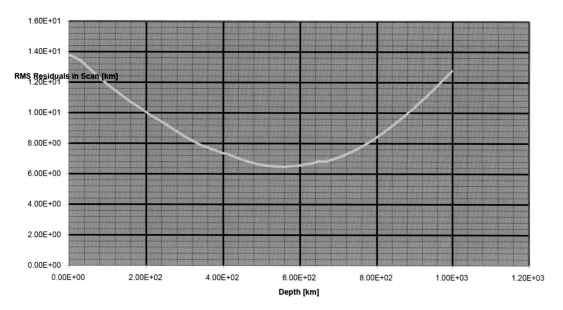

Figure 6.3. Track of tabular scan (epicentre known at 49.78 N, 145.13 E)

The points at which the authors' motivation and work on interpolative tabular scanning with respect to earthquake location problems have been discussed in the literature.[14]

The two types of interpolative scan are resumed here:

- The *first location procedure*, which provides hypocentre depth, H_d, given epicentre coordinates
- The *second location procedure*, giving both epicentre and hypocentre from P-wave first arrivals

Each type, by tabular extension, can provide the take-off angles for each event identified.

Tabular Structure

The table generation routines provide matrix tables of the order of 101×101 to 51×51 ([1],[15] by steering any one of the point-to-point (P2P) ray tracing algorithms, of which there are several, as described in Chapter 4.The main output from this process consists of five objects:

1. P-wave (S-wave) first arrival travel time matrix
2. Corresponding take-off angles matrix
3. Calibration data matrix
4. Error data matrix
5. Error log (a list)

---

[14] GR Daglish & YuP Sizov, *Suggestion for a Rapid Hypocenter Location System using a Tabular Scan*, Conference Paper - Dubrovnik, 2013

[15] GR Daglish, *Creation of Data Structures for Locating Hypocenter Foci by Scanning*, Conference Paper – 33rd General Assembly of the European Seismological Commision, Moscow, 2012.

From 1 to 4, each of the matrices is dimensioned by depth and latitudinal distance. The latitudinal distance, referred to as co-latitude, starts from a conceptual pole at *zero* degrees and extends in the current set (at this instance) to 87 $^{\circ}$.

The contents of the fourth element in this structure are directed at each point in matrices 1 to 3, and the fifth element is of the form of an error log, recording where the chosen P2P algorithm does not achieve the required level of accuracy.

Self-Testing

While the first location procedure was being developed, much attention was paid to determining levels of self-noise within the algorithm. This is exemplified in figures 6.4, 6.5, and 6.6. The first figure (figure 6.4) shows the output of the results of a self-test using one of a set of accuracy-enhancing techniques (2). For each depth point, arrival data is generated. Using this data, a hypocentre scan is performed to relocate each such point. The depth of the hypocentre produced by this scan will correspond to the depth point value, offset by some error or self-noise.

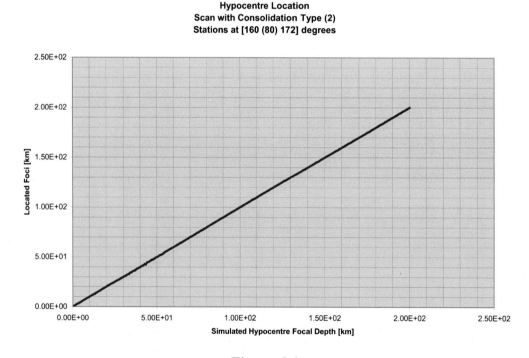

Figure 6.4

The second figure (figure 6.5) shows the self-noise component inherent purely in the granularity of the scan. In this, there are 1,000 data-generating depth points between 0–1000 kilometres. However, the scan proceeds with a granularity of 233 points in the same space.

The third figure (figure 6.6) contrasts the point wise error (self-noise) due to the use of two different scanning techniques. These are discussed in (2) and (5).

Hypocentre Location
Nearest Approach Pattern During Self-test
Granularity of Scan - [1000, 233]

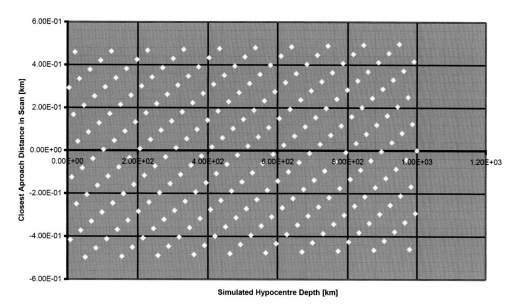

Figure 6.5

Hypocentre Location
Method Comparison
Stations at [30 (40) 72] degrees

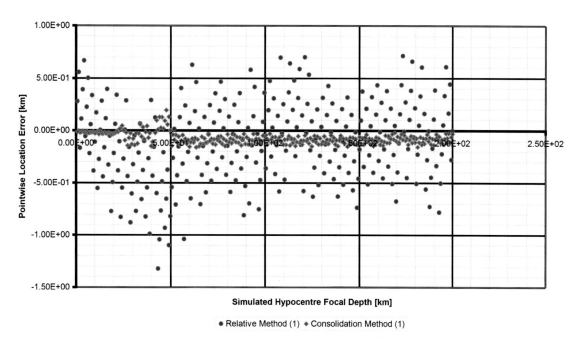

● Relative Method (1) ◆ Consolidation Method (1)

Figure 6.6

However, if the choice of the scanning step length, the generating interval of the tables, and the granularity of the depth-point data coincides, in the sense that they are multiples or submultiples of each other, then zero self-noise levels may ensue. Zero noise levels have been confirmed for such a choice of intervals by many trials.

Hypocentre Location

This is the first location procedure (mentioned in Chapter 4). A description of the associated cost function and the scanning process is resumed here.

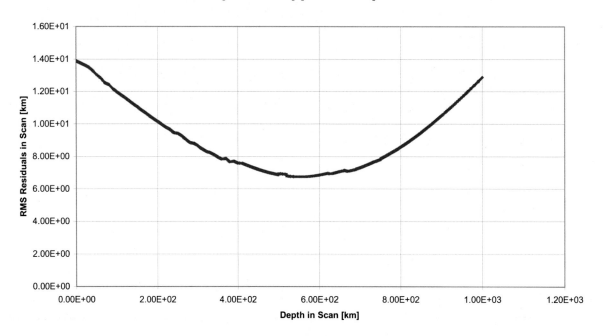

Figure 6.7

The onset timings are made up as follows, from the set T:

$$T \equiv \{t_j\}; \quad j = 0, (n-1)$$

Here the t_j are elements of a universal time running on a continuous time base and represent arrivals of whatever wave species at stations $(j+1)$.. Thus,

$$\bar{t} = \frac{1}{n} \sum t_j$$

We generate a spread of deviations about this mean as

$$\tau_j = \{t_j - \bar{t}\}; \quad j=0, (n-1)$$

This forms the fixed "time template" for the scan. Similarly, using the matrices of travel times generated by the point-to-point spherical ray tracing routines, from any position on the upward/downward scanning trajectory, we can find, by forward interpolation, a set of deviations, indexed as i for $\{v_j\}_i$; $j = 0, (n-1)$, which correspond to the set of known station co-latitudes. In this, *i* indexes a point in the scan, and *j* indexes the set of active seismic stations. For each position in the scan we get

$$RMS_i = \sqrt{\frac{1}{n}\sum_j \left(\tau_j - v_{j,i}\right)^2}$$

It is this set of values that is graphed against depth in Figure 6.7. The infimum is taken to correspond to a hypocentral depth, in this case 555 kilometres. Such a method forms the second example of a "type 1" cost function, described in section V.

Chauvenet's Principle

Chauvenet's principle for the rejection of outliers (11) is installed in the main scanning routine, which implements this first procedure. It is also in the routines that implement the second location procedure.

Examples of the Location Process

Using Wilber 3 (provided to the public by courtesy of IRIS), some data was chosen corresponding to a magnitude 7.3 event in the Kuril Islands. Its epicentre was stated to be at 46.224 N, 150.783 E, $H_d = 112.20$ km, on 19 April 2013. For results, see Annex G.

These results contain several by P wave and one by S wave. The four interpolation methods referred to in the result tables above are:

1. Linear, linear
2. Linear, cubic
3. Cubic, cubic
4. Lagrange, Lagrange

The first method interpolates with depth, and the second, with co-latitude.

Concurrent Epicentre and Hypocentre Location

This is the second location procedure as mentioned at the outset. There appear to be two possible types of *scanning processes* within this context:

1. Method 1 uses a prior estimation of the latitude and longitude of the epicentre of the event.
2. Method 2 does not use a prior estimate of the latitude and longitude of the epicentre of the event.

Each of these two methods has its own cost function—designated type 1 and type 2 indicator systems. We now discuss *method 1* and its corresponding type 1 indicator system.

The co-latitudes of the set of active stations are formed into a set of differences, which set forms a template that is fixed for the entire scan. This is described as the master template.

However, before the formation of this template, the station co-latitudes are transformed relative to the estimated epicentre as pole. Further, the set of co-latitudes is ordered relative to the set of P-wave onset timings, which are sorted into the order of their natural incidences.

The means of estimating the epicentral latitude and longitude are described below.

The next step is the organisation of the set of P-wave first arrival times into a set of differences forming a fixed timing template. Having formed these two templates, the scan commences by:

- Interpolating an entire co-latitudinal row of timings (P-wave first arrivals from the tabular structure referred to above) for a given depth.
- Laterally scanning the fixed timing template along this interpolated row to generate co-latitudes corresponding to its elements by reverse interpolation. By means of a comparison of these co-latitudes with those forming the master template, the position of a local minimum for this type 1 rms indicator is found.
- Repeating the above two processes for each of the set of depth points that form the scan. The smallest local minimum so found is taken to define the hypocentre depth, H_d. The co-latitudes associated with and discovered by this local minimum are used to calculate the epicentral coordinates. This calculation is given below.

To repeat this in plainer language: an actual fixed time template for the lateral scanning procedure consists of a set of differences:

$$\delta t_j = t_j - t_0; \quad j = 0, n-1$$

The base in time for the lateral scan is defined as T_0, and the template is shifted across the depth-interpolated time row as:

$$\tau_i = T_0 + i \cdot \Delta t; \quad i = 0, N-1$$

$$t_j = \tau_i + \delta t_j; \quad j = 0, n-1$$

Here N is the granularity of the scan and $\Delta t = \dfrac{T_{max} - T_0}{N}$. T_0 and T_{max} are the limiting values of the depth-interpolated row. These new t_j are used to reverse interpolate to a set of values for colatitude. At each point, τ_i in the scan, the indicator is formed.

We can be more precise about the methods for comparing a fit to the template under a type 1 cost function. Two methods are given here:

1. We can use the chi-squared variate for goodness of fit as an indicator:

$$\chi^2 = \sum_{i=0}^{n-1} \frac{(O_i - E_i)^2}{E_i}$$

Here the O_i are the co-latitudes extracted via the fixed time template by reverse interpolation. The E_i are the co-latitudes entered into the master template at the start of the scan. Each term making up this variate is stored separately for purposes of the Chauvenet outlier test.

2. The method described above for the first location procedure can be used. In this, we compare residuals, formed relative to a template mean, between an object template and the master template. The differences, $\{d_i\}$, between the residual pairs are used in an rms score:

$$rms = \sqrt{\frac{\sum_{i=0}^{n-1} d_i^2}{n}}$$

And the $\{d_i\}$ are themselves kept to use in a Chauvenet outlier rejection procedure if such has been requested.

The lateral movement of the scan involves a granularity of displacement, which can be set at an arbitrary number of sideways shifts.

Method 2, with its type 2 cost function, proceeds without the use of an estimate of an epicentral location. It does not employ a master template constructed from co-latitudes. Instead it does form a timing template from sorted P-wave first arrivals and uses this in a lateral and downwards scan to derive, by inverse interpolation, co-latitudes from the tabular structure, as stated above. These calculations are used to form radii (as depicted in figure 6.8). These radii are translated and are then subtended from the known station locations. A least-squares calculation for the epicentral co-ordinates ensues. Since the calculation is by least squares and can be well overdetermined, then a "goodness-of-fit" or "self-consistency" assessment can be made for the result. This occurs for each positioning of the fixed timing template in the scan. That location calculation, which provides the best "self-consistency" score, is deemed to indicate the sought-after epicentre and hypocentre pair. This is the type 2 indicator.

This indicator is

$$\sigma = \ln\left(\sum_{i=0}^{n-1} \frac{(v_i - c_i)^2}{c_i}\right)$$

σ is the logarithm of a χ^2 variate and is to be minimised. The v_i are the co-latitudes calculated for the trial epicentre, while the c_i are those formed from the time template. There are n stations.

Both these methods can avail themselves of Chauvenet's principle as a procedure for the rejection of outliers (11) in order to attempt to refine the results. This provides an outer controlling loop on both of the scan types. On detecting an outlier among the set of residuals associated with a given epicentre and hypocentral location, the input data is consolidated to exclude the information associated with the rejected station and the scan recycles. An upper limit is placed for the number of possible recycles, and termination may occur when no further outliers are found prior to this limit being reached.

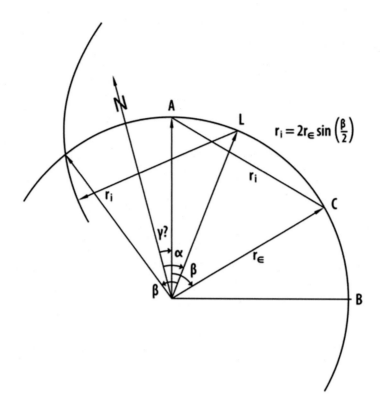

Figure 6.8

The Within-Scan Epicentre Calculation

This description refers to figure 6.8. This figure depicts the geometry associated with a single time point match from a set that corresponds to a local or global minimum of a type 1 or type 2 cost function.

The arc AB represents the latitudinal extent, starting at zero degrees at A, of the P-wave first arrival tables. C represents a point of match for the object time position from the moving time template, now associated with the minimum. This object time position corresponds to a particular station, whose co-latitudinal position is at L. The angle "γ?" is the unknown angle giving the immaterial orientation of the frame of the table to the earth space frame. AC is a chord subtended from the point of match, C, to the "epicentre pole" of the table at A. This chord is then translated and placed with one end at L, which now subtends it as the spherical radius, r_i, of the ith sphere centred at the ith station co-latitude, L.

$$r_i = 2 \cdot r_e \sin\left(\frac{\beta}{2}\right)$$

Once all these radii have been assembled from the matched positions of timings within the template, then the following system of equations is solved for $(x, y, z)_\varepsilon$, which will be a location of the epicentre in the earth space frame:

$$a_i x_\varepsilon + b_i y_\varepsilon + c_i z_\varepsilon = -\frac{1}{2}\left(r_i^2 - r_e^2 - \left(a_i^2 + b_i^2 + c_i^2\right)\right); \quad i = Zero, (n-1). \qquad (0.1)$$

In the above, the $(a, b, c)_i$ are the coordinates of the station positions within the Cartesian earth frame. r_e is a value for the earth radius, while n, (≥ 3), is the number of stations with which the scan is undertaken. The $(x, y, z)_\varepsilon$ is converted to latitude and longitude.

The "type 2" residuals can now be found from this calculation. Each residual is calculated as

$$q_i = \left(\sqrt{\left(x_\varepsilon - a_i\right)^2 + \left(y_\varepsilon - b_i\right)^2 + \left(z_\varepsilon - c_i\right)^2} - r_i\right)^2$$

which gives an rms score of

$$\sqrt{\frac{\sum_{i=0}^{n-1} q_i}{n}}$$

Initial Epicentre Estimation using P-wave First Arrivals

This is at its most effective when proximal stations are used and also when these are within nine degrees of the (unknown) epicentre. However, good results can be obtained up to thirty-two degrees from the epicentre. Obviously, it is not known, a priori, where any group of the proximal (first in receipt of P-wave onsets) stations lie in respect to the epicentre of a given event. Therefore, the calculation has to *evolve*, incorporating more and more data in real time. This is to be the subject of a further paper.

Given a set of stations in the earth frame whose coordinates will allow the least-squares fit to a plane, then such a plane can be fitted as

$$\underline{a} \cdot \underline{x}_i = \rho \qquad (0.2)$$

by solving

$$\begin{bmatrix} \sum x_i^2 & \sum x_i y_i & \sum x_i z_i & -\sum x_i \\ \sum x\, y_i & \sum y_i^2 & \sum y_i z_i & -\sum y_i \\ \sum x\, z_i & \sum z_i y_i & \sum z_i^2 & -\sum z_i \\ -\sum x_i & -\sum y_i & -\sum z_i & -n \end{bmatrix} \cdot \begin{bmatrix} a \\ b \\ c \\ \rho \end{bmatrix} = \begin{bmatrix} 0 \\ 0 \\ 0 \\ -1 \end{bmatrix} \qquad (0.3)$$

The $\underline{a} = \begin{bmatrix} a \\ b \\ c \end{bmatrix}$ are the direction numbers, and the $\underline{x}_i$ are the station coordinates. The $\underline{a}$ are normalised by the factor:

$$d = \sqrt{a^2 + b^2 + c^2}$$

The true value of the radius, ρ, is $\frac{1}{d}$. The whole is rotated so that the radius lies along the $z-axis$ of the earth frame, and the plane is parallel to the xOy plane of the earth frame. Next, the stations are projected either gnomonically or orthogonally onto either of two planes, given by $z = \rho$ or $z = r_e$.

The rotated station coordinates are then used in the planar system following, where the $\{t_i\}$ are derived from the ordered P-wave first arrivals, $\{T_i\}$, processed as $t_i = T_i - T_0$:

$$(x_\varepsilon - x_i)^2 + (y_\varepsilon - y_i)^2 = \overline{c}_0^2 \cdot (t_i - \delta t)^2 ; \quad i = Zero, (n-1)$$

To solve for a rotated epicentre $(x, y)_\varepsilon$ (whose missing $z-ordinate$ can be either ρ or r_e, as has been chosen previously, see previous paragraphs), for $\overline{c}_0$, an in-plano propagation velocity, and δt, an in-plano time to origin from the "lead" station. This epicentre is then rotated back to the original orientation of the plane and the full epicentre estimate, $(x, y, z)_\varepsilon$, (projected onto either the Earth spheroid, or the Earth as sphere), is now the pole for the rotation of the station coordinates. Thus, relative to their rotated frame, the working co-latitudes of the set of stations can be derived.

Examples of Location

The data for each of the following five treatments was derived using Wilber 3 from within the IRIS software system. The event chosen was a magnitude 4.5 event, whose focal depth, H_d, was given as 558.9 km. The latitude/longitude of this event was stated to be 17.8997 S, 181.4731 E. Time and date were 12:19:21 UTC and 3 October 2013.

First Treatment

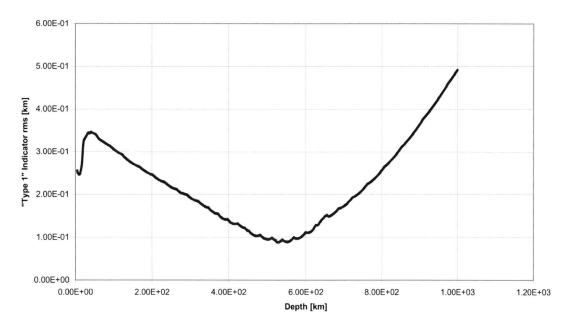

Figure 6.9

The first treatment was a gnomonic projection onto non Earth-tangential plane. Epicentre at 17.913 S, 181.687 E, H_d = 527.375 km. Error in epicentre = 17.87 km; error in Hypocentre focus = 5.642%.

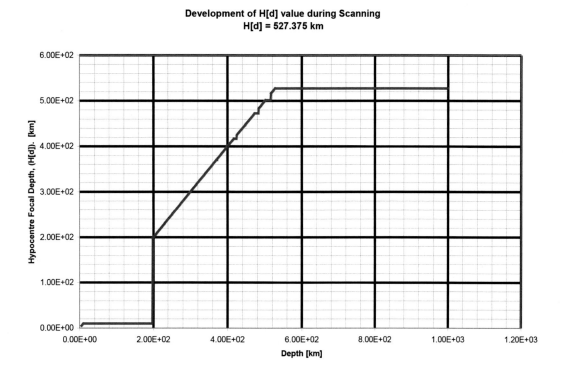

Figure 6.10

Second Treatment

Routines: RQTST 04, HCTST 01, SQTST 01
Gnomonic Projection on Earth-tangent Plane
3 Cycles of Chauvenet's Procedure
Epicentre at 17.977S, 181.685E
H[d] = 541.305 km

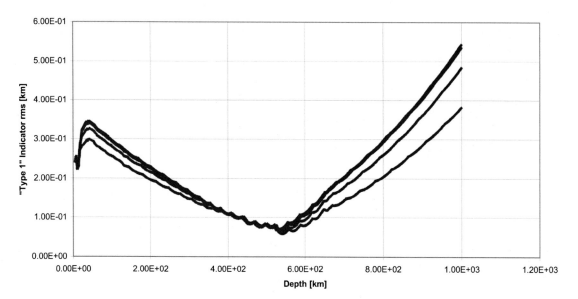

Figure 6.11

The second treatment was gnomonical projection onto earth-tangential plane, with three cycles of Chauvenet's procedure. Epicentre at 17.977 S, 181.685E. H_d = 541.305 km. Error in epicentre = 25.12 km; error in hypocentre focus = 3.15%.

Development of H[d] Values during Scanning
H[d] = 554.24 km

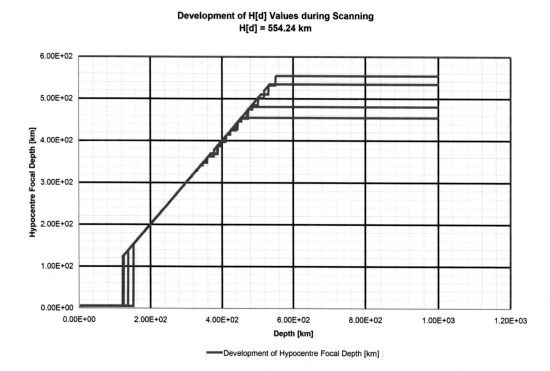

Figure 6.12

Third Treatment

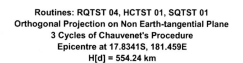

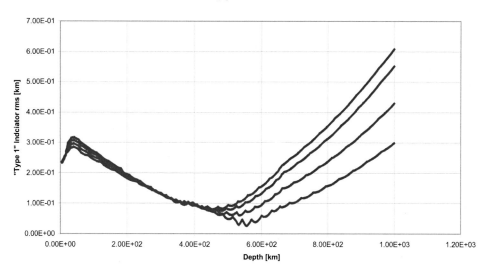

Figure 6.13

The third treatment was orthogonal projection onto non earth-tangential plane, with three cycles of Chauvenet's procedure. Epicentre at 17.834 S, 181.459 E. H_d = 554.24 km. Error in epicentre = 7.47 km; error in hypocentre focus = 0.84%.

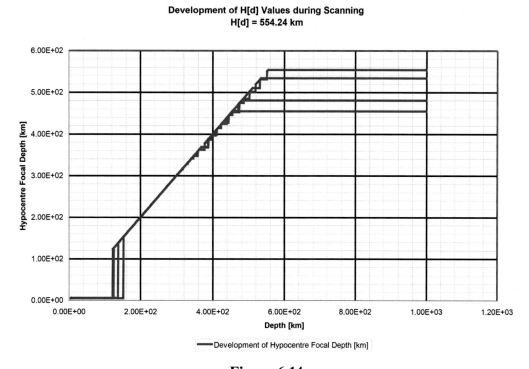

Figure 6.14

Fourth Treatment

Routines: RQTST 04, HCTST 01, SQTST 01
Orthogonal Projection on Earth-tangent Plane
3 Cycles of Chauvenet's Procedure
Epicentre at 17.7893S, 181.4483E
H[d] = 549.265 km

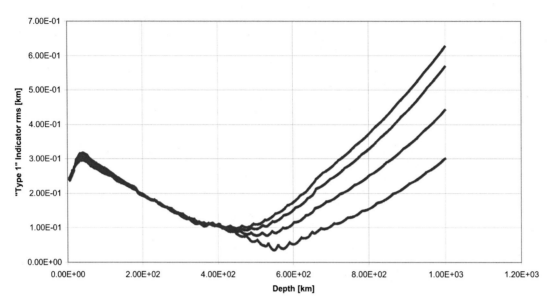

Figure 6.15

The fourth treatment was orthogonal projection onto earth-tangential plane, with three cycles of Chauvenet's procedure. Epicentre at 17.789 S, 181.448 E. H_d = 549.265 km. Error in epicentre = 12.59 km; error in hypocentre focus = 1.73%.

Development of H[d] Values during Scanning
H[d] = 549.265 km

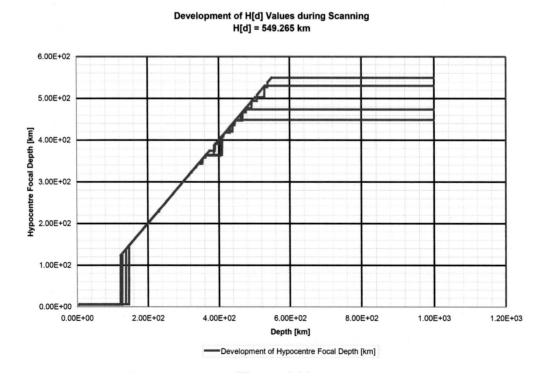

Figure 6.16

Fifth Treatment

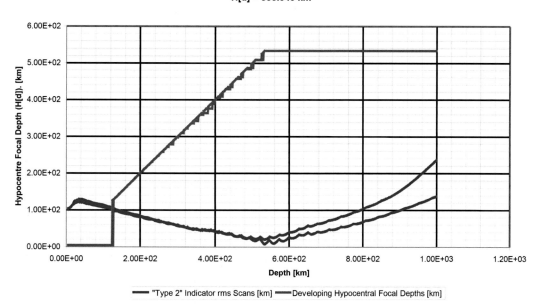

Figure 6.17

The fifth treatment was with no initial estimate of epicentre and with three cycles of Chauvenet's procedure. Epicentre at 17.843 S, 181.459 E. H_d = 533.345 km. Error in epicentre = 9.84 km; error in hypocentre focus = 4.58%.

The fact that the second location procedure uses P-wave first arrivals only would appear to obviate the need for picking, if earthquake early warning (EEW) is the concern, since detection of the initial P-wave onset is quite reliable and well-established.

Therefore in an EEW context, such a method would appear to have some merit. It is able to return the epicentre; hypocentre; and, by extension in the tabular structure, the take-off angles, within approximately four to five seconds of receiving notification of the latest P-wave first arrival in the sequence from which data is to be utilised.

Factors demonstrated in previous trials with the two methods and quoted here, such as:

- low self-noise levels recorded for the first location procedure
- close agreement with IRIS quoted results
- speed of execution

suggest a strong possibility the methods would be "adequate for purpose" as algorithms usable within an EEW context.

Further trials of software structures took place for the second location procedure, which used a reduced form of simulation for real-time conditions by allowing the location system to react to incremental input and to evolve solutions. This location procedure is described in Chapter 7.

Chapter 7

Reduced Simulation of Real-Time Earthquake Localisation – I

Here, "reduced simulation" means the timing data for the localisation process is input in the order that it arises, but without attention to the actual timing at which it appears. This means that there is no attempt made to emulate the buffering that would be required in a truly real-time version of this system.

Figure 7.1 shows the outputs (in red) of the three possible processing paths initially conceived for the concurrent localisation of earthquake epicentres and hypocentres from P-wave first arrivals. This is the situation for Canonical Pathway II, from which Canonical Pathway III was evolved (see Chapter 2).

"Projective" and "non-projective" in the figure imply, respectively, presence or absence of precursor estimates for the epicentre.

The routine SQTST 06 is one step up from the SQTST 01, 02a & 03 routines, in the figure, insofar as it allows time-ordered subsets to be processed incrementally in a reduced simulation of a possible real-time earthquake tracking process.

The final routine, ETTST 10, which is optional, can form a secondary estimation of the hypocentral depth.

In Canonical Pathway III, no attempt is made to find a precursor estimate of the position of the Epicentre, so that it simply follows the left-hand "non-projective" pathway in the figure.

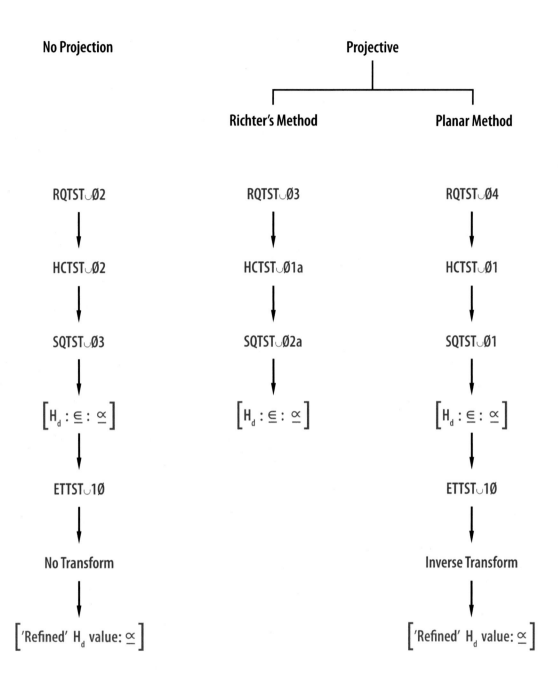

Figure 7.1.

Upgraded version of the concatenated programs ETTST 14.cpp to ETTST 20.cpp, See Annex A

These programs are the main components in Canonical Pathway IV and are formed from the concatenation of the concurrent epicentre and hypocentre localisation routines and the purely hypocentre scanning functions.

The main routines that take part in such trials are

ETTST 14.cpp to ETTST 20.cpp

These routines are characterised by the following features:

1. Cyclically processing incoming subsets of the spreading earthquake data, in the form of P-wave first arrival times
2. Having either a precursor or *no* precursor as prior estimation of the epicentre position, ε.c
3. Using a scan, one of types A to E (see below), to determine a localisation of $\underline{\varepsilon}$ and a corresponding (and tentative) localisation of the hypocentre depth, H_d.
4. Making a further scan (following on from $\underline{\varepsilon}$ in 3) to confirm or moderate the H_d localisation already found
5. Tracking, further, the results as they are progressively produced at each cycle and forming a vector from this information, which is passed to file at each iteration of the routine:
   - Time for each process (mill-time or processor-time)
   - Summation of process times for all subsets
   - Time intervals between data arrivals
   - Summation of all such time intervals
   - Processing time, plus such time interval summation
   - H_d calculated by scan (see 3)
   - Smoothed value of H_d
   - "Error" in determination of $\underline{\varepsilon}$. (As the set of determinations of $\underline{\varepsilon}$ grows, a least-squares calculation forms a running best estimate of $\underline{\varepsilon}$ from the accumulating set $\{\underline{\varepsilon}_i\}$. This is implemented in ETTST 16.cpp, and the calculation is given in chapter 8.)
   - Smoothed value of this "error"
   - Radial discrepancy for the determination of $\underline{\varepsilon}$.
   - Smoothed value for this discrepancy
   - H_d calculated as a secondary moderation to that already found at 3

Following this, the values for each of the four interpolation types in this confirmatory scan are given.

   The tracking, or smoothing of these values (some stationary) is achieved by a choice of:
   - α tracker
   - $\alpha - \beta$ tracker
   - running mean
   - exponentially weighted running mean

With regard to 3, the following types of scan are to be found among the set of ETTST dd.cpp routines previously mentioned:

A) Given *no* precursor estimation, a least-squares overdetermined location calculation is made to determine the coordinates for $\underline{\varepsilon}$ within the earth space frame. Each such calculation gives rise to a self-consistency measure. As these measures are generated for calculations that progress across the range of co-latitudes associated with a set of source points at increasing depth, their minimum is recorded. This is assumed to correspond to the sought-after values for $\underline{\varepsilon}$ and H_d..

B) Also, given *no* precursor estimate for $\underline{\varepsilon}$,, a statistic is generated, again as the scan traverses a range of co-latitudes and depth points. A minimum value of this statistic is taken to indicate the locality of the required $\underline{\varepsilon}$ and H_d pair. This statistic is:

$$\sigma_n = \sum_1^{n-1}\left\{\left(Unity - \frac{S_{i-1}}{S_i}\right)\cdot T_i - \delta t_{i-1}\right\} \qquad (0.4)$$

This is explained as follows: A subset of this statistic set is generated across a scan over a set of co-latitudes for a specific depth point. The minimum value within this set is taken to define that set of radii to be used to make a least-squares calculation for $\underline{\varepsilon}$., at that depth. These minima are accrued for the range of depth points considered, and their minimum is taken to indicate the existence of the required source localisation.

In the above expression, S_{i-1} represents the ray path length immediately preceding that defined by S_i. T_i is the travel time for the ray path, S_i, and δt_{i-1} represents the given time difference between P arrivals at the ith and $(i-1)^{th}$ stations. These terms, in the summation for σ_n, are determined from the consideration

$$\frac{(Path\ length)_i - (Path\ length)_{i-1}}{Mean\ Velocity\ for\ (Path\ length)_i} = Difference\ in\ Arrival\ Times$$

Using the above notation, this gives

$$\frac{(S_i - S_{i-1})}{\left(\frac{S_i}{T_i}\right)} - \delta t_{i-1} = \left(Unity - \frac{S_{i-1}}{S_i}\right)\cdot T_i - \delta t_{i-1}$$

The assumption is that σ_n will correspond to a minimum, when it most closely reflects the truth of the hypothesis:

$$\left(Unity - \frac{S_{i-1}}{S_i}\right)\cdot T_i - \delta t_{i-1} \to Zero$$

C) The third kind of scan simply uses the precursor $\underline{\varepsilon}_0$ to generate a template consisting of the co-latitude of each receiving station, relative to itself as pole. Then the elements of a scanned version of the template formed from the set of time differences between successive arrivals (by inverse interpolation from the tables of travel times) are compared with those of the precursor template. Closest agreement is taken to indicate the depth of H_d and the position of $\underline{\varepsilon}$.

D) Given a precursor estimate, a scan can be made (within the co-latitude and depth range considered) using the scanned co-latitude template to form timings to compare with the set of time differences of the input P arrivals. When these sets of timings are compared, the closest agreement is taken to indicate the sought-after ε and H_d.

(A variant to the scan types C and D is to use scan type A to scan in depth, employing the minimal positions found for each co-latitudinal scan. This will generate a set of independent measures, the minimum of which will indicate H_d.)

In symbolic terms, scan type C and scan type D may be described in the following manner. For C, symbol definitions are:

$$\underline{\varepsilon}_0 \rightarrow \{C_i\}; \ \{\tau_i\} \rightarrow \{\Delta T_i\}$$

$$\text{scan: } \{\Delta T_i\} \rightarrow \{t_i\}$$

$$\{t_i\} \rightarrow \{Q_i\} \quad \text{(inverse interpolative look-up)}$$

$$\text{minof } \left[\{Q_i\} \text{ cf } \{C_i\}\right] \Rightarrow (\underline{\varepsilon} \wedge H_d)$$

For D, they are:

$$\underline{\varepsilon}_0 \rightarrow \{C_i\}; \ \{\tau_i\} \rightarrow \{\Delta T_i\}$$

$$\text{scan: } \{C_i\} \rightarrow \{c_i\}$$

$$\{c_i\} \rightarrow \{t_i\} \quad \text{(direct interpolative look-up)}$$

$$\text{minof } \left[\{\delta t_i\} \text{ cf } \{\Delta T_i\}\right] \Rightarrow (\underline{\varepsilon} \wedge H_d)$$

Here: $\{\tau_i\}$ gives the P-wave first arrival times at the set of stations..

$\underline{\varepsilon}_0$ is a precursor estimate of the coordinates of the epicentre.

$\{C_i\}$ are the set of relative co-latitudes based on $\underline{\varepsilon}_0$.

$\{c_i\}$ are the sideways shifted set of co-latitudes during scanning.

$\{t_i\}$ are the resultant travel times by the direct interpolative extraction by $\{c_i\}$ from the 101×101 or 51×51 travel-time matrix.

$\{\Delta T_i\}$ is the arrival time first differences from the given input of P-wave initial times of onset.

$\{\delta t_i\}$ are the resultant set of time differences from the interpolated $\{t_i\}$ during the scanning process.

$\{Q_i\}$ gives the relative co-latitudes generated by inverse interpolation on the travel time tables to be compared with $\{C_i\}$.

118

E) The fifth type of scan is a scan for H_d, performed only when given a relatively secure knowledge of $\underline{\varepsilon}$ and the station co-latitude template generated from it. This has been fully described in chapter 6.

In summary, we have table 7.1. The letters in the table correspond to the scanning types given above.

Table 7.1. Scanning routine facilities and structure

Routine	No precursor	No precursor	Precursor	Scan for H_d
ETTST 14	A	~	~	E
ETTST 15	~	B	~	E
ETTST 16	A	~	~	E
ETTST 17	~	~	C	E
ETTST 18	~	~	D	E
ETTST 19	~	~	No Scan	E
ETTST 20	~	~	IRIS Data for $\underline{\varepsilon}$	E

The precursors are calculated in four possible ways (table 7.2), given a plane of best fit through the set of active sensors chosen.

Table 7.2. Modes to generate precursors

Plane position Projection Type	In Situ	Earth -tangential
Orthogonal	×	×
Gnomonic	×	×

The position of the projective plane can be chosen as the fitted plane position itself (in situ) or removed to be tangential to the earth. The stations can be projected onto either of these plane positions, gnomonically or orthogonally. Once the stations have been projected onto the chosen plane, a precursor for $\underline{\varepsilon}$ is calculated by solving the system:

$$\left(x - a_i\right)^2 + \left(y - b_i\right)^2 = \overline{c}_0^2 \left(t_i - \delta t\right)^2; \quad i = Zero, \, n-1 \qquad (0.5)$$

for $\underline{x}$, δt, and $\overline{c}_0$, within the plane. Here, n is the number of chosen stations; $\underline{x} = \begin{pmatrix} x \\ y \end{pmatrix}$ is the position of the precursor of $\underline{\varepsilon}$ on the plane; $\underline{a}_i = \begin{pmatrix} a_i \\ b_i \end{pmatrix}$ are the position vectors of the projected stations inscribed on the plane; t_i are the arrivals at the stations relative to some lead-station arrival; and δt is the time to source as seen from the chosen lead station. It is a negative value relative to the time origin "located" at the lead station. $\overline{c}_0$ is, in some sense, an averaged propagation velocity, in plano.

Once the position of $\underline{\varepsilon}$ is established in the plane, it is projected, from the centre onto the geosphere. Using the inverse transform, which took the plane radius vector to lie on the +ve z-axis of the Earth space frame, the vector $\underline{\varepsilon}$ is transformed back to the position in which it should lie in the untransformed state. A transform is then found which carries it to the +ve Z-axis, as well as taking the station locations with it. From this, the relative co-latitudinal positions can be calculated with $\underline{\varepsilon}$ as pole.

All output is tracked and placed in a generic set of files:

ABTrackedData ETdd nn qq.txt

dd is the serial number of the routine

nn is the earthquake identifier

qq is the number of active stations

This output in all cases can serve as input to the routine

ABResult.cpp

which applies an:

- α – tracker
- $\alpha\beta$ – tracker
- exponentially weighted mean

as a set of three passes serially output to

ABTrackedDataOut.txt

These three passes form three separate blocks of information in this latter file. The file can then be passed to Excel for interpretation and graphical display. The results from the given earthquakes can be stored as a set of event-specific files.

Chapter 8

Reduced Simulation of Real-Time Earthquake Localisation – II
(The Concatenated Program "ETTST 16.cpp")

This chapter describes the workings and output of the program ETTST 16.cpp. The program is a concatenated version of all that has gone before. It suggests a means by which confidence limits may be placed around the epicentre and hypocentre values that are evolved, concurrently and in real time, on the detection of an increasing number of P-wave first arrivals sequentially provided by the present algorithm, (reduced simulation).

As described in previous chapters, this algorithm is table driven, in that it uses extraction by an interpolative tabular scanning process to deliver its results. For this purpose, a set of three main tables are provided—a table of travel times for rays originating from a graduated set of depth points to a given set of co-latitudes, a table for a set of take-off angles corresponding to the travel times, and a set of calibrating tables to correct numerical error inherent in the table-generating process itself. These tables are generated by any from a set of point-to-point (P2P) ray tracers parameterised by any from a set of radial earth velocity models (PREM, iasp91, ak135).

The production of the confidence limits is concomitant on considering that each value (in other words, the discovered epicentre and hypocentre) is stationary and subject to perturbation by error.

In brief, the error is considered, ultimately, to be normal. Therefore, normal theory is used, as in Chauvenet's test to screen the production of individual localisations for outlying data inputs. Subsequently, it is used to monitor for outliers in the set of localisations themselves.

Use is then made of the t variate (Student's-t) to dynamically establish confidence limits, of varying levels of significance, about regressions on the set of localisations as this set increases in real time.

The upshot is that the algorithm can produce successive localisations of the earthquake as data input (P-wave onset times) arises and, at the same time, monitor the accuracy and integrity of the solutions.

Introduction

The aim of this chapter is to demonstrate that:

- The algorithm structure described below will produce results in close accord with the received values.
- This overall structure is capable of assessing the error (to given levels of significance) in this output.

The authors feel that the algorithm shall be employed as a rapid and lightweight front end for possible earthquake early warning (EEW) systems. By using incoming data to rapidly evolve solutions for

- epicentre and hypocentre localisation
- take-off angle

for events in real time, there can evolve, from an initial small set of onsets, localisations of increasing accuracy. It must be emphasised that only the timings of P-wave first arrivals are initially required from the activated stations. This being one of the clearest observations an automatic energy onset detection system can make, would make it a prime candidate for inclusion in the overall software architecture of a centralised facility supporting EEW.

The build-up to this algorithm is given elsewhere, where it is shown how the algorithm appears to branch out from the main progression of localisation techniques, being table-driven and dependent on precalculated tables for use in an interpolative tabular scan. As stated in Chapter 4, these tables are up to five-fold:

1. Travel times for rays from each depth point in a set of depth points to a set of specific co-latitudes
2. Take-off angles for each ray so treated
3. Calibration data for later interpolation to correct error due to the ray tracer's occasional non- convergence
4. Accuracy data monitoring the degree of convergence of each ray traced
5. Path length of each ray

These five tables are also accompanied by a log of non-convergence, which lists those points at which the P2P ray tracers did not achieve the required accuracy.

The timing for the primary (see below), and now more complex, algorithm is about 0.7 seconds per station for a set of four to eight stations, about 1.3 seconds per station for a set of up to thirty stations, and about 2.15 seconds per station for a set of up to forty-five stations—running on a 3.2 GHz processor.

Data Structures and Modus Operandi

The theory alluded to above is implemented in the routine

ETTST 16.cpp

This routine will evolve a set of sequential solutions or localisations as it accrues data from elements in the network surrounding the event. Each localisation is accompanied by the following vector of information:

1. Number of stations in localisation
2. Ordinal number for localisation
3. Time in main algorithm(s)
4. Running total of 3

5. Delta of total processing times for main algorithm, secondary algorithm, (see below at point 28), and statistical processing: $(T_{i+1} - T_i)$

6. Running total of 5

7. $3 + 6$

8. χ^2 value

9. Significance flag (-1 or +1)

10. Level of probability

11. Threshold for rms comparison on Calculation of $\underline{\varepsilon}$ (epicentre coordinates)

12. Actual rms from final calculation of $\underline{\varepsilon}$

13. Radial discrepancy (as percentage)

14. Primary hypocentre, H_d

15. Tracked $1°$ H_d values

16. $\delta\varepsilon = |\underline{\varepsilon} - \underline{\varepsilon}_I|$ (calculated epicentre, $\underline{\varepsilon}$, and reference epicentre, $\underline{\varepsilon}_I$, from IRIS)

17. Tracked $\delta\varepsilon$ value

18. Radial discrepancy, $\delta\rho$, as percentage (again) $= \dfrac{|\rho - R_\varepsilon|}{R_\varepsilon} \times 100$. ρ is the derived earth radius, while

    R_ε is an earth radius, local to the latitude given at 20.

19. Tracked $\delta\rho$ value

20. Latitude

21. Longitude

22. ρ (calculated earth radius) – Then, 20, 21, and 22 are for the calculated $\underline{\varepsilon}$, as in 16.

23. x_I

24. y_I

25. z_I – Here, 23, 24, and 25 are IRIS Cartesian coordinates in earth space frame for $\underline{\varepsilon}_I$ as reference.

26. Latitude

27. Longitude for IRIS epicentre – $\underline{\varepsilon}_I$

28. Consensus on secondary hypocentre, H_d, from 29, 30, 31, and 32, all produced by the secondary algorithm, which runs these distinct interpolation processes in tandem

29. $2°$ H_d by linear/linear interpolation

30. $2°$ H_d by linear/cubic interpolation

31. $2°$ H_d by cubic/Lagrange interpolation

32. $2°$ H_d by Lagrange/Lagrange interpolation

33. θ, an angle between the current Tracked version of $\underline{\varepsilon}$ and $\underline{\varepsilon}_I$ or the prior $\underline{\varepsilon}$.

34. Length of the spherical arc subtended by θ

35. A similar arc distance to the above, at 33 and 34, but between $(\underline{\varepsilon}_{i+1} - \underline{\varepsilon}_i)$

36. Upper confidence limits for the arc at 35

37. Lower confidence limits for the arc at 35

38. Average or "best estimate" for 35, $\langle \Delta\underline{\varepsilon} \rangle$

39. Upper bounds for the confidence interval on 38

40. Lower bounds for the confidence interval on 38, $\langle \Delta\underline{\varepsilon} \rangle$

41. Primary H_d for this localisation (stated again)

42. Upper bounds for the confidence interval on 41

43. Lower bounds for the confidence interval on 41, $1°$ H_d

44. Average or "best estimate" for 41, $\langle H_d \rangle$

45. Upper bounds for the confidence interval on 44
46. Lower bounds for the confidence interval on 44, $\langle H_d \rangle$

This set of vectors, produced sequentially, is destined for the output file:

ABTrackedData.txt

As stated, the main point for the algorithm is to produce an epicentre/hypocentre pair, together with an error assessment. In the above vector definition, we can see that the output fulfils this. The program ETTST 16.cpp incorporates two scanning algorithms in tandem:

1. A primary algorithm produces the primary epicentre/hypocentre pair.
2. A subsequent algorithm takes the epicentre from the aforementioned pair and, knowing this, performs a secondary scan for hypocentre localisation. The two hypocentres thus found tend to agree closely.

This program performs a restricted emulation of the output of an earthquake event insofar as it takes in data (P-wave onset times) from stations (with corresponding station coordinates) in strict sequence. It does not, however, attempt to emulate the real-time buffering that would be necessary if the input stuck to the actual time differences between the incoming data items.

The file that contains the input data for the entire process associated with one earthquake is

HypoCentreData.txt.

The output from ETTST 16.cpp, apart from a narrative to the monitor, goes in the following directions (these data tables apply to a single unique earthquake):

HypoCentreResultSQ.txt. Contains all output relating to all the localisations performed, including take-off angles.

StationPolarTransformSQ.txt. A record of the input subsets and traces of each scan.

Combinations02Sequence.txt. A set of traces of the localisations made on all subsets for this earthquake.

ABTrackedData.txt. The set of data vectors for the localisations realised for each subset, structured as above.

EpiCentrePolar04.txt. The set of epicentre localisations for each subset in this earthquake.

The above output is created from the following input, ancillary to the main input table, given above:

EpicentrePolar03.txt. Contains the received value for the epicentre of this earthquake.

StationLatLong.txt. Dimensions of depth and co-latitude for the following matrix tables.

PSArrivalTabulation00.txt. Matrix of P-wave travel times (seconds) from source at depth to a surface co-latitude.

TakeOffAngleGrid.txt. Matrix of take-off angles corresponding to the P-wave rays in the above Matrix.

RayPathDistance.txt. Matrix of ray path lengths from source to target surface co-latitude.

Overview of Software Action

The processing of each localisation proceeds in five main phases (for symbols, see above):

1. The next epicentre/hypocentre pair $(\underline{\varepsilon}, H_d)$ is produced by the primary algorithm (optionally using a Chauvenet procedure internally to the process).

2. The $\underline{\varepsilon}$ and the H_d are tracked, generating the $\delta\varepsilon$. This also generates the parameters for the secondary algorithm—in other words, the set of relative co-latitudes for each active station based on the position of the primary $\underline{\varepsilon}$.

3. The production of the secondary and consensus H_d from the four interpolation regimes.

4. The application of the external Chauvenet procedure to the series $\{\delta\varepsilon_i\}$ and $\{H_{d_i}\}$.

5. Incorporating the results of 4 (in other words, taking into account the temporarily eliminated elements of these two latter series), confidence limits are generated for:

   - last value of $\delta\varepsilon$
   - best estimate $\langle \delta\varepsilon \rangle$
   - last value of H_d
   - best estimate $\langle H_d \rangle$

The application of the Chauvenet test (11) takes place in two situations at two levels:

- In *phase 1*, to screen and eliminate, if necessary, data used for the primary algorithm, within that algorithm
- In *phases 4 and 5*, to screen and eliminate, if necessary, elements of the series of localised

$$\{\delta\varepsilon_i\}; \quad \{H_{d\,i}\}$$

and to use the resulting map of eliminations in generating

$$\langle \delta\varepsilon \rangle; \quad \langle H_d \rangle$$

The means whereby the best estimates are produced and how the Chauvenet procedure is implemented are described in upcoming sections.

The system is tested by taking seven known earthquakes and passing their data through the routine ETTST 16.cpp (see above).

Summaries of these known earthquakes are given in table 8.1.

The data used for these trials is taken, using Wilber 3 (supported by IRIS). Their parameters are tabulated in table 8.1.

Table 8.1. Earthquake parameters

Number	Region	Magnitude (M_w)	Epicentre Lat/ Long (degrees)	H_d (km)
1	West of North American Coast	5.4	41.7136N/126.8446W	9.9
2	Fiji Islands region	4.5	17.8997S/ 178.5269W	558.9
3	Just East of Honshu	5.4	35.621N/ 140.6862E	36.0
4	South Mid Atlantic Trench	5.2	47.1861S/13.4304W	10.0
5	Celebes Sea	4.1	3.9999N/ 123.3981E	495.1
6	Yunnan (China)	6.1	27.446N/103.427E	10.0
7	Santa Cruz Islands	4.7	12.23985S/167.135E	259.3

Results

Although there are seven earthquakes in the process, only the results of the seventh will be fully presented, since space does not allow for all seven. However, the results for the entire set can be made available on request.All subsequent figures, 8.1 to 8.6, apply to earthquake 7, Santa Cruz Islands.

The sequences of localisations found by this algorithm generally are characterised by an initial "erratic" or "transient" period, followed by a phase when the output is stabilised around a relatively constant set of values that is shown in figure 8.3. The figure displays the output from the first localisation using four stations to the twelfth localisation, using fifteen inputs.

Figure 8.1 shows a comparison between the rms for residuals arising from the final epicentre calculations and threshold values formed from a percentage of the mean of the radii also involved in the same calculation. It can be seen that the first points (5 to 15) are more erratic than those subsequent to point 15, which remain on a more regular "trajectory". This characterises the transient phase for this set of localisation

Zones 5 to 15 in figures 8.1 and 8.2 are complementary, with the erratic behaviour in 8.3 corresponding to the disturbed zone in 8.2 Hereafter, most of the graphs begin at about the level of fifteen inputs (in other words, the twelfth localisation). In fact, the calculations establishing the confidence intervals (see below) do not output until they have accumulated ten points of data to form mean and sigma values. Because of this initial wildness of the output, the confidence intervals tend to be wider at the start.

The effect of the Chauvenet rejections (see also section dealing with "Chauvenet Test for Outliers") is discussed in the next section.

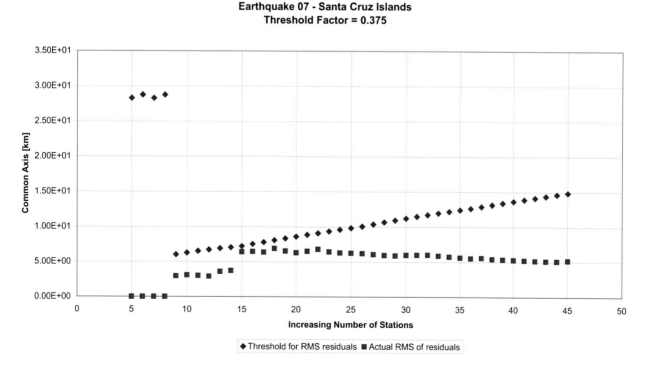

Figure 8.1.

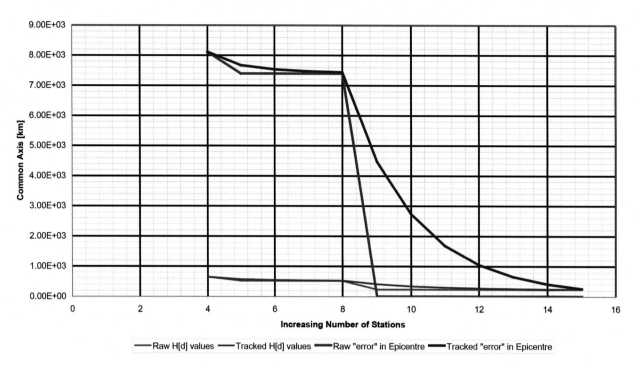

Figure 8.2.

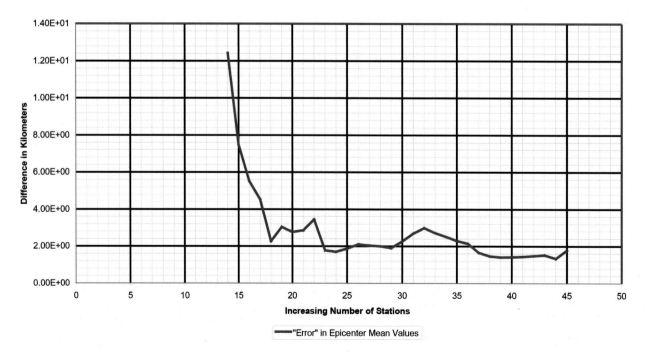

Figure 8.3.

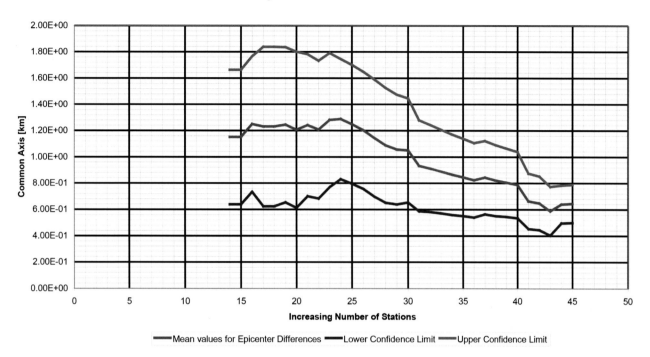

Figure 8.4. Arc distance between successive values for epicentres, $\langle \Delta \underline{\varepsilon} \rangle$

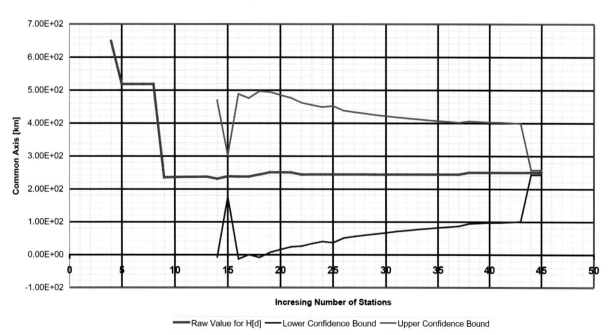

Figure 8.5. Series of raw H_d **values**

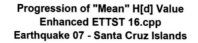

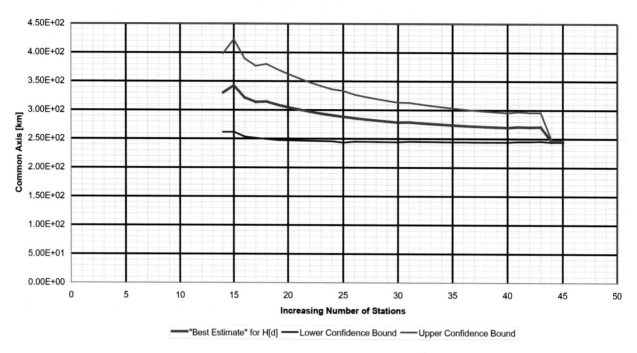

Figure 8.6. Series of "best estimates" of H_d

Discussion of Results

The main features to note are:

- The hypocentre was delivered at 251.0 ±8.0 kilometres
- The epicentre was delivered at 12.2266 S, 176.144 E.
- The distance between the delivered epicentre and the received value was $1.78 \pm 7.89_{10} - 01$ kilometres. (This delivered epicentre was estimated by the method given in the next sections.)
- The confidence intervals are seen to severely contract when the number of rejected localisations peaked at the end of the run (see figures 8.5 and 8.6).

These values are very close to the received values, and their confidence intervals are narrow.

It can be seen that the evolution of the confidence intervals follows more or less closely the gradually increasing number of rejection of localisations by the Chauvenet test. This is apparent when comparing the list of rejected stations, below, with the graph in figure 8.6.

The results arising from the processing of earthquake 7 (Santa Cruz Islands), given here, would appear to show that this algorithm—which concurrently delivers epicentres and hypocentres by interpolative tabular scanning—is capable of monitoring the accuracy of its output reliably and in real time.

Best Estimate of Epicentre from the Accumulating Set $\{\underline{\varepsilon}_i\}$

This forms a calculation for a point on a sphere, $\underline{\alpha}$, as "centroid" to a set of other points, $\underline{x}_i$, on a sphere. We have

$$\alpha x_i + \beta y_i + \gamma z_i = r_\varepsilon^2 \cos \vartheta_i$$

Here, $\underline{\alpha} = \begin{pmatrix} \alpha \\ \beta \\ \gamma \end{pmatrix}$ and $\underline{x}_i = \begin{pmatrix} x \\ y \\ z \end{pmatrix}_i$. r_ε is the local Earth radius. ϑ_i is the angle between $\underline{\alpha}$ and $\underline{x}_i$, at Earth

centre. Considering the arc length subtended by ϑ_i, namely S_i, we have

$$S_i = r_e \vartheta_i = r_e \cos^{-1}\left(\frac{1}{r_e^2} \cdot \left(\alpha x_i + \beta y_i + \gamma z_i \right) \right)$$

We should seek to minimise the sum:

$$\frac{F_n}{r_e^2} = \frac{1}{r_e^2} \sum_i S_i^2 = \sum_i \vartheta_i^2 = \sum_i \left\{ \cos^{-1}\left(\frac{1}{r_e^2} (\underline{\alpha} \cdot \underline{x}_i) \right) \right\}^2$$

on behalf of the elements of $\underline{\alpha}$.

$$f_i = \cos^{-1}\left(\frac{1}{r_e^2} (\underline{\alpha} \cdot \underline{x}_i) \right); \quad \underline{f} = \left[f_0 f_1 f_2 \ldots f_{n-1} \right]$$

$$\frac{\partial f_i}{\partial \alpha, \beta, \gamma} = -\frac{1}{r_e^2} \cdot \frac{x_i, y_i, z_i}{\sqrt{1 - \dfrac{(\underline{\alpha} \cdot \underline{x}_i)^2}{r_e^4}}}$$

Let $J_{\underline{f}}$ be the Jacobean of $\underline{f}$, and then, applying the Gauss–Newton iterative method, we have

$$\underline{\alpha}_{i+1} = \underline{\alpha}_i + \left(J_{\underline{f}}^T \cdot J_{\underline{f}} \right)^{-1} \cdot J_{\underline{f}}^T \cdot \underline{f}^T \bigg|_i$$

This is considered to have converged when the modulus of the adjustment vector

$$-\underline{h}\big|_i = \left(J_{\underline{f}}^T \cdot J_{\underline{f}} \right)^{-1} \cdot J_{\underline{f}}^T \cdot \underline{f}^T \bigg|_i$$

is less than a given small value, for example one metre.

However, if the convergence is not satisfactory, then it can be accelerated by including the Hessian matrix of the cost function in the calculation of the adjustment vector

$$\underline{\alpha}_{i+1} = \underline{\alpha}_i + \left(J_{\underline{f}}^T \cdot J_{\underline{f}} + \sum_{j=0}^{n-1} f_j \cdot H_{f_j} \right)^{-1} \cdot J_{\underline{f}}^T \cdot \underline{f}^T \bigg|_i$$

Here the Hessian matrices, H_{f_j}, are constructed as

$$-\frac{1}{R_\varepsilon^4}\begin{bmatrix} \dfrac{x_0^2\cdot\left(\underline{\alpha}\cdot\underline{x}_j\right)}{\left(\sqrt{A_j}\right)^3} & \dfrac{x_0 x_1\cdot\left(\underline{\alpha}\cdot\underline{x}_j\right)}{\left(\sqrt{A_j}\right)^3} & \dfrac{x_0 x_2\cdot\left(\underline{\alpha}\cdot\underline{x}_j\right)}{\left(\sqrt{A_j}\right)^3} \\[4mm] \dfrac{x_1 x_0\cdot\left(\underline{\alpha}\cdot\underline{x}_j\right)}{\left(\sqrt{A_j}\right)^3} & \dfrac{x_1^2\cdot\left(\underline{\alpha}\cdot\underline{x}_j\right)}{\left(\sqrt{A_j}\right)^3} & \dfrac{x_1 x_2\cdot\left(\underline{\alpha}\cdot\underline{x}_j\right)}{\left(\sqrt{A_j}\right)^3} \\[4mm] \dfrac{x_2 x_0\cdot\left(\underline{\alpha}\cdot\underline{x}_j\right)}{\left(\sqrt{A_j}\right)^3} & \dfrac{x_2 x_1\cdot\left(\underline{\alpha}\cdot\underline{x}_j\right)}{\left(\sqrt{A_j}\right)^3} & \dfrac{x_2^2\cdot\left(\underline{\alpha}\cdot\underline{x}_j\right)}{\left(\sqrt{A_j}\right)^3} \end{bmatrix},$$

and

$$A_j = Unity - \left(\frac{\underline{\alpha}\cdot\underline{x}_j}{R_\varepsilon^2}\right)^2$$

The Chauvenet test for Outliers

This simple test of significance consists of comparing a normalised variate against a value taken from a precalculated table. This table establishes outlying limits of normal distribution according to the order of the set from which the candidate-normalised variate is generated. If the normalised variate is greater than the table value, then it may be judged an outlier.

The value from the table is indexed by n, the order of the set from among whose elements outliers are required to be identified. This is a single tails test, and the normalised variate builds up as

$$\frac{\left|h_i - \overline{h}\right|}{\sigma_h}$$

Here, the set is $H = \{h_i\}$ with $\overline{h}$, the mean, and σ_h^2, the variance. h_i is a general set element to be screened for disassociation from H.

A more in-depth way of describing this process of disassociation, which is Chauvenet's test, is to state as follows:

"If the population from which the sample is drawn is normal then

$$P_{>|Z_i|} = 1 - \sqrt{\frac{2}{\pi}}\int_0^{z_i}\exp\left(-\frac{z^2}{2}\right)dz; \quad z_i = \frac{h_i - \overline{h}}{\sigma_h}$$

represents that proportion of the population that is equal to or exceeds the value h_i. This proportion can be forced to be:

$$n \cdot \left(1 - \sqrt{\frac{2}{\pi}} \int_0^{z_i} \exp\left(-\frac{z^2}{2} \right) \cdot dz \right) \leq \frac{1}{2} \text{ or } \frac{1}{q_0}$$

so that, now h_i represents a margin of probability at and beyond which only smaller and smaller non-integral sample elements can exist, then we can decide to call $|h_i|$ and all smaller values outliers."

We can say that the expected or probable relative frequency for the occurrence of the value $\langle h_i \rangle$ is at the most

$$\frac{1}{n \cdot q_0}.$$

Further, we can solve for z_0 in the inequality below for a set of n elements and parameterised by values of q_0:

$$\left(1 - \sqrt{\frac{2}{\pi}} \int_0^{z_0} \exp\left(-\frac{z^2}{2} \right) dz \right) \leq \frac{1}{n \cdot q_0}$$

n is the number of elements in the sample set $H = \{h_i\}$, and $q_0 \geq 2$ is a set of arbitrary values. Then

$$z_0 = \frac{h_i - \overline{h}}{\sigma_h}; \quad h_i = \sigma_h \cdot z_0 + \overline{h}$$

defines outliers at the relative frequency level $\dfrac{1}{n \cdot q_0}$ ..

List of Rejections by Chauvenet's Procedure

This section lists those localisations that have been rejected as outliers by applying the Chauvenet principle. The first line, for instance, reads at localisation 11. The results of localisations 1 and 5 were excluded from the calculations leading to the best estimates of whatever variables and their corresponding confidence intervals.

Such variables would be $\langle H_d \rangle$ and $\langle \Delta \underline{\varepsilon} \rangle$ (see section above for symbols).

(Note: Localisation 1 corresponds to the first data set of order 4 and so on.)

```
11: 1 5: Total 2
12: 1 5 11 12: Total 4
13: 1 5 11: Total 3
```

```
14:  1 5 11:  Total 3
15:  1 5 11 14:  Total 4
16:  1 5 11 14:  Total 4
17:  1 5 11 14:  Total 4
18:  1 5 11 14:  Total 4
19:  1 5 11 14:  Total 4
20:  1 5 11 14:  Total 4
21:  1 5 11 14:  Total 4
22:  1 5 11 14:  Total 4
23:  1 5 11 14:  Total 4
24:  1 5 11 14:  Total 4
25:  1 5 11 14:  Total 4
26:  1 5 11 14:  Total 4
27:  1 5 11 14:  Total 4
28:  1 5 9 11 14:  Total 5
29:  1 5 9 11 14:  Total 5
30:  1 5 9 11 14:  Total 5
31:  1 5 9 11 14:  Total 5
32:  1 5 9 11 14:  Total 5
33:  1 5 9 11 14:  Total 5
34:  1 5 9 11 14:  Total 5
35:  1 5 9 11 14:  Total 5
36:  1 5 9 11 14:  Total 5
37:  1 5 9 11 14:  Total 5
38:  1 5 9 10 11 14 19:  Total 7
39:  1 5 9 10 11 14 19:  Total 7
40:  1 5 9 10 11 12 14 19:  Total 8

41:  1 2 3 4 5 9 10 11 12 14 19:  Total 11

42:  1 2 3 4 5 9 10 11 12 14 19:  Total 11
```

Best Estimation and Confidence Intervals

We are dealing with a set of values, say hypocentre depths, collectively H_d, which correspond to a set of ordinal numbers, integers, which have no error. The readings of H_d are deemed constant, it could be the same value, and are therefore only perturbed by error in the data and the self-noise in the algorithm. The overall error distribution affecting the values H_d is assumed normal, invoking the central limit theorem.

Therefore, a line of regression is fitted to the row of perturbed H_d. For clarity we say that the set of hypocentre depths is denoted by $H = \{h_i\}$, whose order is n, the index of the last localisation.

The line of regression will have been found as

$$\langle h_i \rangle = a_n + b_n \cdot i; \quad i \in [1, n]$$

This regression will be performed anew for each localisation as n increases.

A best estimate of H_d from the set H, representing a value assumed stationary, would be

$$\langle h \rangle = \frac{1}{n} \sum_i h_i$$

while

$$s_h^2 = \frac{\sum_i \varepsilon_i^2}{v}; \quad v = n - 2$$

where

$$\varepsilon_i = h_i - \left(a_n + b_n \cdot i\right)$$

Then

$$s_{\bar{h}}^2 = \frac{s_h^2}{n}$$

by which the confidence interval is given as

$$I = \left[\langle h \rangle - t \cdot s_{\bar{h}}, \; \langle h \rangle + t \cdot s_{\bar{h}}\right]$$

Here the t-variate is indexed from tables by the degree of freedom, v, and the level of significance (probability level) required. In these tests, a level of 5 per cent is chosen. Thus, the value $\bar{h}$ has a 95 per cent chance of finding itself within the interval I.

The above refers to the best estimate and its confidence interval. However, we also would like to put a confidence interval about the latest entry to the set H. We write

$$s_{h_i}^2 = s_{h_{\bar{i}}}^2 + s_h^2$$

$$s_{h_i}^2 = s_h^2 \left[Unity + \frac{1}{n} + \frac{\left(h_n - \bar{h}\right)^2}{\sum_i \left(h_i - \bar{h}\right)^2} \right]$$

The required confidence interval is

$$I = \left[\langle h_n \rangle - t \cdot s_{h_n}, \; \langle h_n \rangle + t \cdot s_{h_n}\right]; \quad \langle h_n \rangle = a_n + b_n \cdot n$$

Chapter 9

Concurrent Epicentre and Hypocentre Localisation using a Spheroidal Earth Model

We have proposed fast algorithms for the concurrent localisation of earthquake epicentres and hypocentres in real time. In tests, a "reduced simulation" regime is followed, in which these algorithms absorb P-wave first arrival timings sequentially into the object data set, in the order in which the expanding wave field impinged on the sensors (seismographs) in the network considered. Most localisation algorithms at some point apply a correction for the earth's ellipticity. However, the bulk of the calculation is done assuming a spherical earth—with the correction added on as an afterthought.

In this chapter, we will broach a proposal for implanting the localisation algorithm directly within the earth as an oblate spheroid (the spheroid). This would obviate the need for a "tag-on" correction.

The processing of each localisation proceeds in five main phases:

1. The next epicentre/hypocentre pair, $\left(\underline{\varepsilon}, H_d \right)$, is produced by the primary algorithm as its dataset absorbs a timing from the next seismograph in the sequence.
2. The $\underline{\varepsilon}$ and the H_d are tracked, generating the $\delta\varepsilon$ (arc distances between successive localisations of the epicentre). This also generates the parameters for the secondary algorithm—in other words, the set of relative co-latitudes for each active station based on the position of the primary $\underline{\varepsilon}$.
3. There is a secondary and consensus H_d which is produced from the four independent interpolation regimes that have been described at Chapter 4.
4. The external Chauvenet procedure is applied to the series $\left\{ \delta\varepsilon_i \right\}$ and $\left\{ H_{d_i} \right\}$.
5. The results of 4 are incorporated (in other words, the temporarily eliminated elements of this sequence are taken into account), and confidence limits are generated for:
    *   last value of H_d
    *   best estimate $\left\langle H_d \right\rangle$

The application of the Chauvenet test [16] takes place in two situations at two levels:

*   In *phase 1*, to screen and eliminate, if necessary, data used for the primary algorithm, within that algorithm
*   In *phases 4 and 5*, to screen and eliminate, if necessary, elements of the series of localised

$$\left\{ \delta\varepsilon_i \right\}; \quad \left\{ H_{d\,i} \right\}$$

---

[16]   AM Neville & B Kennedy, *Basic Statistical Methods for Enigineers and Scientists*, Intertext Books, 1964.

and to use the resulting map of eliminations in generating

$$\langle \delta\varepsilon \rangle; \ \langle H_d \rangle$$

The means whereby the best estimates are produced and the Chauvenet procedure is implemented are described at the end of chapter 8.

The system is tested by taking seven known earthquakes and passing their data through the routine QLoc00.cpp.

For summaries of these known earthquakes, see table 9.1.

The data used for these trials is taken, using Wilber 3 (supported by IRIS). Their parameters are tabulated in table 9.1.

Table 9.1. Earthquake parameters

Number	Region	Magnitude [M(w)]	Epicenter Lat/long (degrees)	H[d] (km)
1	West of North American Coast	5.4	41.7136N/ 126.844W	9.9
2	Fiji Islands Region	4.5	17.8997S/ 178.5269W	558.9
3	Just East of Honshu	5.4	35.621N/ 140.6862E	36.0
4	South Mid-Atlantic Trench	5.2	47.186S/ 13.430W	10.0
5	Celebes Sea	4.1	3.9999N/ 123.3981E	495.1
6	Yunnan (China)	6.1	27.446N/ 103.42E	10.0
7	Santa Cruz Islands	4.7	12.2398S/ 167.135E	259.3

This section deals with the location procedures in finer detail as mentioned at the outset. We begin with the scanning process. Although there appear to be two possible ways of scanning within this context

1. a scan that makes no use of any prior estimation and provides, simultaneously, values for epicentral coordinates and the hypocentral focal depth (here the *primary process*)
2. a scan that makes use of a prior estimate of the coordinates of the event epicentre and provides a value for the hypocentral focal depth only (here the *secondary process*[17])

space dictates that only the first, and more important, type will be dealt with in detail here.

---

[17] GR Daglish & YuP SIzov, *System Suggestions for Hypocenter Location using Tabular Data with some Illustrative Results*, Journal of Computer Engineering and Information, October 2013, 96 – 111, World Academic Publishing (WAP)

The co-latitudes and longitudes of this set of active stations are formed into a set of Cartesian co-ordinates within the earth space frame. These are to be used later in the localisation calculation, which is used to derive the epicentral position.

The next step is the organisation of these to P-wave first arrival times into a set of differences forming a fixed timing template.

Having formed these two sets of information, the scan commences by:

- Interpolating an entire co-latitudinal row of timings (P-wave first arrivals from the tabular structure referred to above) for the next depth point reached in the scan
- Laterally scanning the fixed timing template along this interpolated row to generate co-latitudes corresponding to its elements by reverse interpolation
- Repeating the above two processes for each of the set of depth points that form the scan, where the smallest local minimum of the indicators found is taken to define the epicentral co-ordinates and the hypocentre depth, H_d

To repeat this in plainer language, an actual fixed time template for the lateral scanning procedure consists of a set of differences:

$$\delta t_j = t_j - t_0; \quad j = 0, n-1$$

The base in time for the lateral scan is defined as T_0, and the template is shifted across the depth-interpolated time row as

$$\tau_i = T_0 + i \cdot \Delta t; \quad i = 0, N-1$$

$$t_j = \tau_i + \delta t_j; \quad j = 0, n-1$$

Here N is the granularity of the scan and

$$\Delta t = \frac{T_{max} - T_0}{N}$$

T_0 and T_{max} are the limiting values of the depth-interpolated row. These new t_j are used to reverse interpolate to a set of values for co-latitude. At each point, τ_i, in the scan the indicator is formed. This indicator is

$$\sigma = \ln \left(\sum_{i=0}^{n-1} \frac{(v_i - c_i)^2}{c_i} \right)$$

Then σ is the logarithm of a χ^2 variate and is to be minimised. The v_i are the co-latitudes calculated for the trial epicentre, while the c_i are those formed from the time template. There are n stations. The calculation for the trial epicentre is as follows:

At each point in the scan, the co-latitudes c_i are used to form radii (as depicted in figure 9.1). These radii are translated and then subtended from the known station locations. A least-squares calculation for the trial epicentral coordinates ensues.

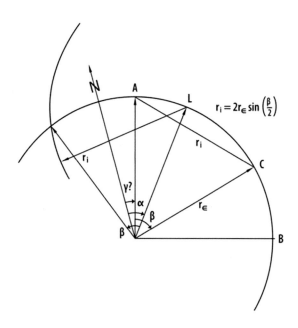

Figure 9.1. Geometrical basis for within-scan epicentre location

In the figure, the arc AB represents the latitudinal extent, starting at zero degrees at A, of the P-wave first arrival tables. C represents a point of match for the object time position from the moving time template, now associated with a possible minimum. The angle "γ?" is the unknown angle giving the immaterial orientation of the frame of the table to the earth space frame. AC is a chord subtended from the point of match, C, to the epicentre pole of the table at A. This chord is then translated and placed with one end at L, which now subtends it as the spherical radius, r_i, of the ith sphere centred at the ith station co-latitude, L.

$$r_i = 2r \cdot \sin\left(\frac{\beta}{2}\right)$$

Once all these radii have been assembled from the matched positions of timings within the template, then the following system of equations is solved for $(x, y, z)_\varepsilon$, which will be a possible location for the epicentre within the earth space frame:

$$a_i x_\varepsilon + b_i y_\varepsilon + c_i z_\varepsilon = -\frac{1}{2}\left(r_i^2 - r_\varepsilon^2 - \left(a_i^2 + b_i^2 + c_i^2\right)\right); \quad i = Zero, (n-1)$$

In the above, the $(a, b, c)_i$ are the coordinates of the station positions within the earth frame. r_ε is a value for the earth radius, while n, (≥ 3), is the number of stations with which the scan is undertaken. The $(x, y, z)_\varepsilon$ is converted to latitude and longitude.

The above equations represent the incorporation of Earth spherical geometry, while the following development represents the incorporation of Earth spheroidal geometry into the localisation process.

We now take into account "N" simultaneous spherical equations centred on the station positions (a_i, b_i, c_i) and constrain the least-squares estimate of their intersection to lie on the spheroid:

$$\frac{x^2}{r_e^2} + \frac{y^2}{r_e^2} + \frac{z^2}{r_p^2} = 1$$

This becomes

$$x^2 + y^2 + \frac{z^2}{\epsilon^2} = r_e^2$$

where

$$r_e = Earth\ Equatorial\ Radius$$

$$r_p = Earth\ Polar\ Radius$$

$$\epsilon = \frac{r_p}{r_e}$$

On subtraction from each of the sphere equations:

$$2(x \cdot a_i) - z^2(1 - \frac{1}{\epsilon^2}) + (r_i^2 - r_e^2) - \| a_i \|^2 = F_i(x)$$

where

$a_i = (a,\ b,\ c)_i;\ \ i \in [1, N]$ – *Sensor coordinates subtending the r_i in Earth Space Frame*
$x = (x, y, z)$ – *Epicentre coordinates on Spheroid*

$$\frac{\partial F_i}{\partial x} = 2a_i$$

$$\frac{\partial F_i}{\partial y} = 2b_i$$

$$\frac{\partial F_i}{\partial z} = 2\left(c_i - z\left(1 - \frac{1}{\epsilon^2} \right) \right)$$

This non-linear scheme can be solved by the reduced Gauss–Newton iteration given by

$$x_{i+1} = x_i - \left(J^T \cdot J \right)^{-1} \cdot J^T \cdot F \mid_i$$

Here J is the Jacobean for the function vector F evaluated at the ith step of the iteration.

The following figures show the progression of values achieved by both these versions of algorithm that use tables of P2P travel times as the successive P-wave first arrivals timings are added in.

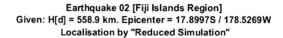

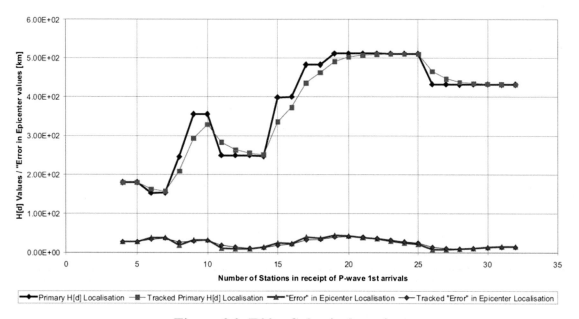

Figure 9.2. E02 – Spherical version

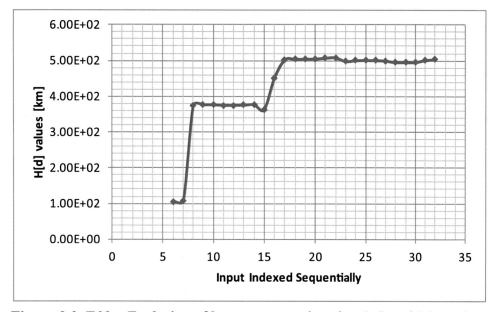

Figure 9.3. E02 – Evolution of hypocentre estimation (spheroidal version)

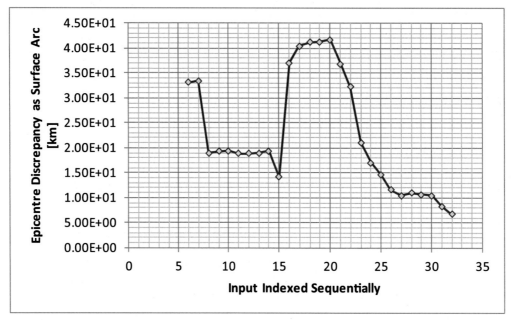

Figure 9.4. E02 – Discrepancy in epicentre (spheroidal version)

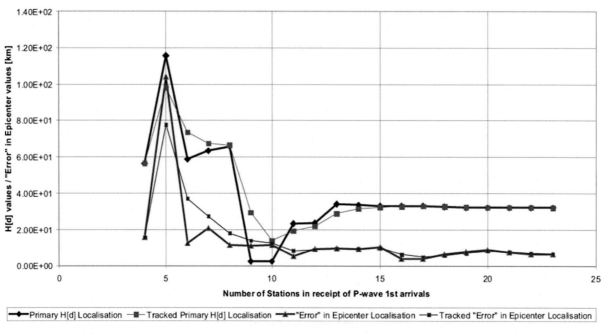

Figure 9.5. E03 – Spherical version

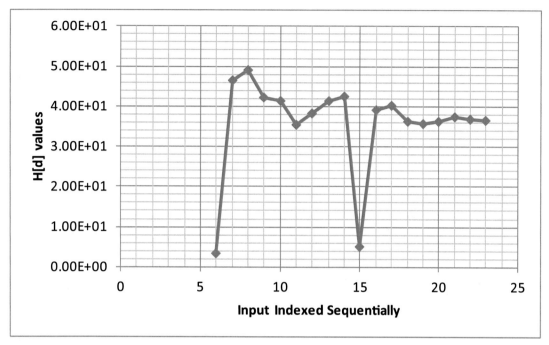

Figure 9.6. E03 – Evolution of hypocentre estimation (spheroidal version)

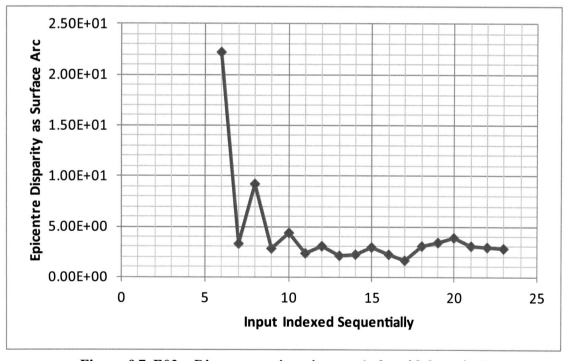

Figure 9.7. E03 – Discrepancy in epicentre (spheroidal version)

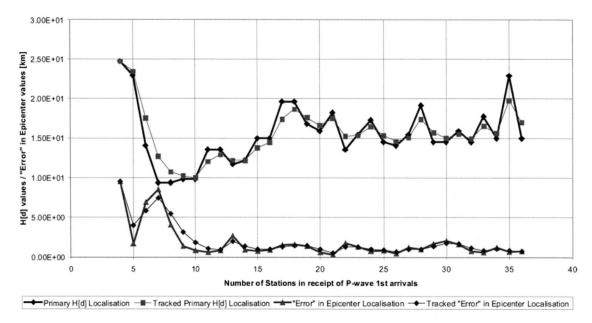

Figure 9.8. E04 – Spherical version

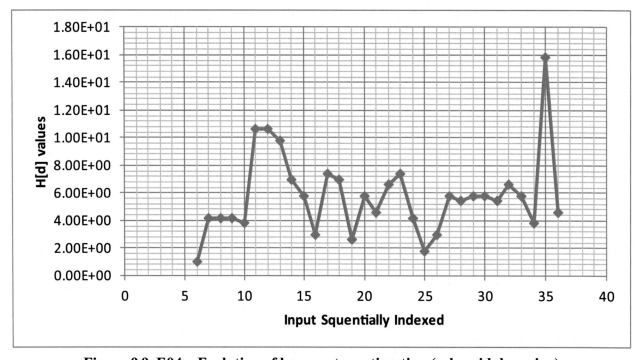

Figure 9.9. E04 – Evolution of hypocentre estimation (spheroidal version)

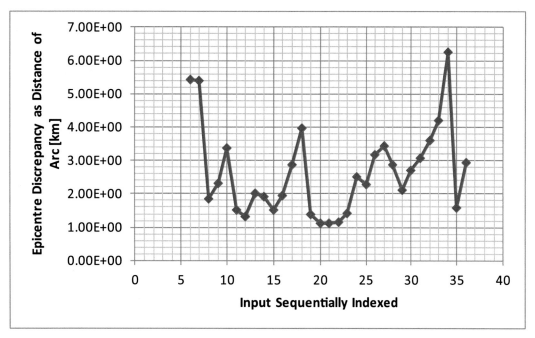

Figure 9.10. E04 – Discrepancy in epicentre (spheroidal version)

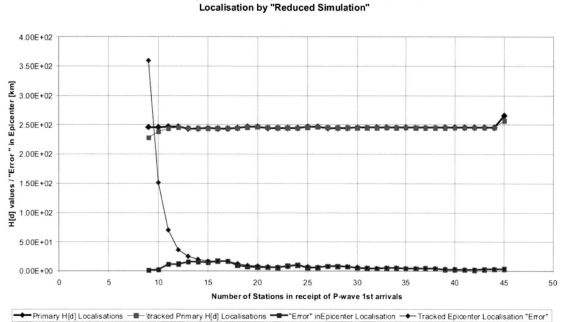

Figure 9.11. E07 – Spherical version

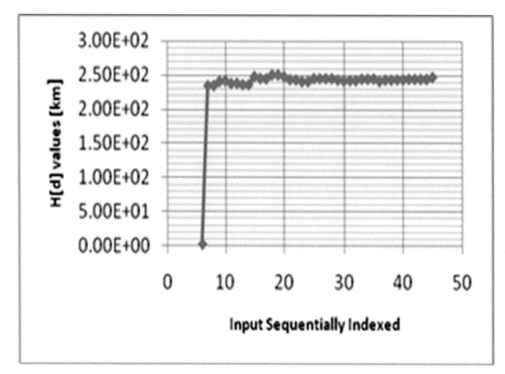

Figure 9.12. E07 – Evolution of hypocentre estimation (spheroidal version)

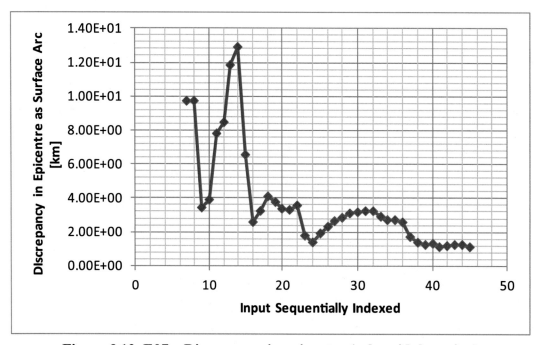

Figure 9.13. E07 – Discrepancy in epicentre (spheroidal version)

Figures 9.1 to 9.13 form a comparison between the tabular scans that (a) follow a spherical earth model and (b) follow a spheroidal earth model over earthquakes E02, E03, E04, and E07. Qualitative observations are made:

- In both cases, initial instability occurs in the trajectories and dies out as more data accrues
- With regard to epicentral discrepancy, type (b) gives final values smaller than type (a).

- The iterative nature of the epicentre solution in type (b) would need to have its convergence criteria tightened.
- The solutions for H(d) in both cases are in close agreement with each other and also with the values provided by table 9.1.

Estimation of Epicentral Coordinates from (a) Time Differences within Individual Seismograms and (b) Direct Arrivals of Love and Rayleigh Waves

For two separate arrivals at a given point in space, originating from the same source and having travelled with different velocities, we can write

$$v_0 t = v_1 \left(t + \Delta t \right)$$

since each has passed over the same distance. Here, $v_0 t$ corresponds to the distance covered by that arrival with the greater velocity. And $v_1 \left(t + \Delta t \right)$ is the distance covered by the arrival with the lesser velocity. These two distances are the same.

The time to origin for the first arrival would be

$$t = \frac{v_1}{v_0 - v_1} \cdot \Delta t$$

Thus, knowing Δt, the time difference between the two arrivals, by observations on the given seismogram, we may write

$$s = v_0 t = \frac{v_0 v_1}{v_0 - v_1} \cdot \Delta t$$

or

$$s = v_1 \left(t + \Delta t \right) = \left(\frac{v_1}{v_0 - v_1} + Unity \right) \cdot v_1 \Delta t$$

Then, s is the distance travelled from the source event by both wave species. From this, we can construct the following diagram.

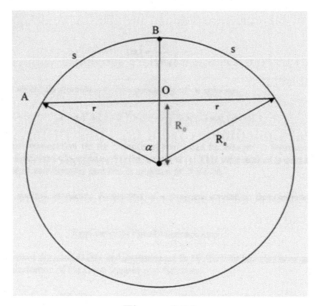

Figure 9.14

A and B are positions of the source and the receiving station respectively. Here, R_e is an effective earth radius. In this sagittal diagram we find

$$\alpha R_e = s \implies \alpha = \frac{s}{R_e}$$

$$r = R_e \sin\alpha; \quad R_0 = R_e \cos\alpha$$

In three-space, the coordinates of the point O, in the same frame as the coordinates of point B, are

$$a_0 = \frac{R_0}{R_e}\cdot a; \quad b_0 = \frac{R_0}{R_e}\cdot b; \quad c_0 = \frac{R_0}{R_e}\cdot c,$$

where the position of O is (a_0, b_0, c_0) and the position of B (the station) is (a, b, c). The point O, therefore, is the centre of a sphere:

$$(x-a_0)^2 + (y-b_0)^2 + (z-c_0)^2 = r^2,$$

whose surface will include the source position at A. Thus, for many stations and correspondingly observable Δt_i, we get a set:

$$\{\alpha_i\} = \left\{\frac{s_i}{R_e}\right\}$$

from which we might generate a system consisting of n spheres:

$$\left(x-a_i\right)^2 + \left(y-b_i\right)^2 + \left(z-c_i\right)^2 = r_i^2; \quad i = 0,(n-1)$$

whose mutual intersection (in the absence of error) can be thought to represent the position of epicentral coordinates for the source at A. This intersection is calculated as a linearised least-squares problem in program DLTST 00.

To the algorithm depicted in this and implemented in **DLTST 00** has also been added a direct minimisation of the least-squares cost function:

$$S_n = \sum_{i=0}^{i=n-1} \left(\sqrt{u_i} - r_i\right)^2; \quad u_i = \left(x-a_i\right)^2 + \left(y-b_i\right)^2 + \left(z-c_i\right)^2$$

which uses the output from the linearised version of the equations (1.6) as a starting value for its iterations. This is a Gauss–Newton iterative algorithm, which will, firstly, use Jacobean matrices but will pull in a Hessian matrix if the main iterations are going astray.

With direct estimation from Rayleigh and Love waves, using slightly different notation from that used at the outset, we can say:

The development for the simple spherical shell or lamina is given here as a conceptual equation system— in particular

$$\underline{x} \cdot \underline{a_i} = R^2 \cos\left(\frac{\overline{c}_0\left(t_i - \Delta t\right)}{R}\right); \quad i = 0,(n-1)$$

where

$\underline{x}$ is the location of the Seismic Emission in cartesians.

$\underline{a_i}$ is the location of the ith sensor in cartesians.

R is the radius of a spherical shell.

t_i is the ith arrival time relative to the lead sensor time.

Δt is the time-to-origin, i.e. the arrival time at the lead sensor.

$\overline{c}_0$ is, in some sense, an averaged wave propagation velocity.

Then the solution vector for this system is

$$\underline{s} = \begin{bmatrix} \underline{x} \\ \Delta t \\ \overline{c}_0 \end{bmatrix}$$

The Cartesians referred to here are coordinates within a space frame whose origin coincides with the centre of the object sphere. Thus, in (1.7)

$$\|\underline{x}\| \approx R; \quad \|\underline{a_i}\| \approx R.$$

This fact (or supposition) allows the following derivation:

An arc length is given by an expression such as $R\theta$, where R is the radius of the sphere or circle in question. Therefore, considering the sphere on whose surface emissions are being transmitted and removing the coordinate system origin to the centre of this sphere we get

$$R\cos^{-1}\left(\frac{xa_i + yb_i + zc_i}{\|\underline{x}\| \cdot \|\underline{a}_i\|}\right) = \overline{c}_0\left(t_i - \Delta t\right); \quad \iota \in \left[0, (n-1)\right],$$

But :

$$\|\underline{x}\| \cdot \|\underline{a}_i\| = R^2.$$

So :

$$R\cos^{-1}\left(\frac{xa_i + yb_i + zc_i}{R^2}\right) = \overline{c}_0\left(t_i - \Delta t\right).$$

Transposing functions gives

$$xa_i + yb_i + zc_i = R^2 \cos\left(\frac{\overline{c}_0\left(t_i - \Delta t\right)}{R}\right); \quad i \in \left[0, (n-1)\right],$$

as above. This system is then solved for $\left(x, y, z, \Delta t, \overline{c}_0\right)$.

To fix $\underline{x}$, using this system, $n \geq 5$. It is notable that the above system is in the form of a set of planes, the lengths of whose normals to the origin are time dependent and oscillate as cosine functions.

The restitution of $\underline{x}$ can be considered to represent the point where all n planes mutually intersect and where, indeed, the point of intersection lies on the spherical surface of radius R.

The solution of system (1.7) proceeds in two stages. An initial approximation to the solution vector is found by a linear least-squares scanning method using a "spider" (as a gauge of the curvature) moving over a region of a surface defined in a time/velocity space. See Annex J and consider for all subsequent versions of this direct technique with a spherical constraint.

Using this initial value, a Gauss–Newton "descent" method is employed to find the root of the above system, considered as a highly non-linear least-squares cost function.

Further considerations on the direct method for finding epicentre coordinates are described in the remainder of this chapter.

We shall resume and develop a second type of equation for these purposes. Consider an arc length Γ on a great circle inscribed on a sphere. Consider also the chord it subtends, of length s = 2d. Then, if we wish to transform one into the other, in terms of length, to get Γ from s, we should multiply by the ratio

$$\rho = \frac{\alpha R_e}{2R_e \sin\left(\dfrac{\alpha}{2}\right)}$$

Thus

$$\Gamma = s \cdot \frac{\alpha}{2\sin\left(\dfrac{\alpha}{2}\right)}$$

Here, α is the angle subtended by the arc and chord pair at the centre of the great circle, while R_e is the sphere (earth) radius. To get the chord length from the arc length, we must multiply by the ratio

$$\tilde{\rho} = \frac{1}{\rho} = \frac{2\sin\left(\dfrac{\alpha}{2}\right)}{\alpha}$$

Considering that we are dynamically observing events originating on or within a sphere and are aware of such events by virtue of the transmitted surface wave phenomena they emit, we can write, for the arc length Γ_i

$$\Gamma_i = \overline{c}_0\left(t_i - \delta t\right)$$

Here,

$\overline{c}_0$ is the propagation velocity of the surface wave in question;

δt is the time to origin—in other words, the time taken for the energy to reach this sensor from the position of the event;

t_i is the elapsed time formed from subtracting the time at which the energy reached the lead sensor from the time it impinged upon the ith sensor. Thus, t_0 is, in fact, *Zero*.

So, if we have many sensors we may have many arcs traversed in the like manner. Now

$$\frac{\alpha_i}{2} = \frac{\overline{c}_0\left(t_i - \delta t\right)}{2R_e}$$

Therefore to convert the observed set of arcs, $\left\{\Gamma_i\right\}$, to chords, $\left\{s_i\right\}$, we can operate as follows:

$$s_i = \Gamma_i \cdot \frac{2\sin\left(\dfrac{\alpha}{2}\right)}{\alpha} = \Gamma_i \cdot \frac{2\sin\left(\dfrac{\overline{c}_0\left(t_i - \delta t\right)}{2R_e}\right)}{\dfrac{\overline{c}_0\left(t_i - \delta t\right)}{R_e}}$$

substituting for gamma

$$s_i = \overline{c}_0\left(t_i - \delta t\right) \cdot \frac{2R_e\sin\left(\dfrac{\overline{c}_0\left(t_i - \delta t\right)}{2R_e}\right)}{\overline{c}_0\left(t - \delta t\right)}$$

so

$$s_i = 2R_e\sin\left(\frac{\overline{c}_0\left(t_i - \delta t\right)}{2R_e}\right)$$

If we then allow chords to be subtended between the sensor points on the surface of our sphere and the unknown position of the source, we can write the following system of equations for locating the approximate epicentre of this source:

$$(x-a_i)^2 + (y-b_i)^2 + (z-c_i)^2 = 4R_e^2 \sin^2\left(\frac{\overline{c}_0(t_i - \delta t)}{2R_e}\right); \quad i = 0,(n-1).$$

with further development this system may be solved for the following:
$\underline{x} = (x, y, z)$ the position of source in Cartesians,
δt the time to origin, and
$\overline{c}_0$ the propagation velocity.
We note, for the solution of this system, we must also add the constraint that the Cartesians must lie on the surface of the sphere:

$$x^2 + y^2 + z^2 = R_e^2$$

We can, therefore, add or subtract this spherical locus from the above system, giving, say:

$$\underline{x} \cdot \underline{a}_i = R_e\left(\frac{1}{2} - 2\sin^2\left(\frac{\overline{c}_0(t_i - \delta t)}{2R_e}\right)\right)$$

Here we have subtracted, and it is now possible to solve this system with great facility using the abbreviated form of Gauss–Newton, which springs from the quadratic approximation method. We have

$$-\underline{h}_i = \left(J^T J\right)^{-1} J^T \cdot \underline{f}\Big|_i$$

Here,
$\{\underline{h}_i\}$ represents the set of successive step lengths in the solution path.
J is the Jacobean for the function vector $\underline{f}$, evaluated at the ith point in the iteration.

We must also remark that the spherical locus should be added to, or subtracted from, the first system, for direct calculation of epicentre coordinates, namely

$$x \cdot a_i + y \cdot b_i + z \cdot c_i = R_e^2 \cos\left(\frac{\overline{c}(t_i - \delta t)}{R_e}\right); \quad i = 0,(n-1).$$

This would lead to, on subtraction

$$\underline{x} \cdot (\underline{a}_i - \underline{x}) = R_e^2\left(\cos\left(\frac{\overline{c}_0(t_i - \delta t)}{R_e}\right) - Unity\right); \quad i = 0,(n-1).$$

This also can be solved with great facility using the Gauss–Newton method, as mentioned in the main text or as sketched above. Or the full quadratic approximation method may be deployed as

$$\underline{x}_{i+1} = \underline{x}_i - \left(J^T J + \sum_{j=0}^{n-1} f_j H_{f_j} \right)^{-1} \cdot J^T \underline{f} \bigg|_i$$

Here,

J is the Jacobean of the vector of functions $\underline{f}$ evaluated at iteration point i. H_{f_j} is the Hessian of the function f_j evaluated at iteration point i.

Chapter 10

Earthquake Localisation and the Elliptic Correction

In this chapter, we will broach a proposal for implanting the earth's spheroidal geometry into scanning localisation algorithms directly. This implantation can act either within a tabular scan delivering a concurrent localisation of epicentres and hypocentres (see chapter 9) or within a purely hypocentral scan, which needs knowledge of the epicentre. This would obviate the need for a tag-on correction. Output from this spheroidal scanning option is included with comment. The consequence for the computational power that would be needed to support such a set of calculations is also treated.

Introduction

The need for an ellipticity correction in the localisation of earthquake epicentres and hypocentres was mooted by Gutenberg and Richter in 1933 (1). Analyses by Jeffreys[18] and Bullen[19] followed in 1935 and 1937 respectively, with tables of corrections to the travel times of P-waves and S-waves.

Since 2008, with the Sichuan earthquake in China, the authors (having initially been encouraged by Dr Lei Hou of High Performance Computer Laboratory, Shanghai University) have developed a family of localisation algorithms. The algorithms are based on the interpolative scanning of tables of ray path travel times and other associated tables generated by any from a set of point-to-point (P2P) ray tracers, also in conjunction with specific earth velocity models.

These slimmed down algorithms locate epicentres and hypocentres concurrently and are designed to do so in real time[20]. This work has been carried out in parallel with, and independently of, more comprehensive investigations carried out in France by Anthony Lomax.

In this chapter, we will demonstrate the mathematics that has led to the consideration of a possible localisation algorithm using the geometry of the spheroid directly within its P2P ray tracer components.

Specifically, the algorithm to be presented currently represents a hypocentre scan based on an approximate knowledge of the epicentre, determined (on a spherical or spheroidal earth model) by any of a plethora of methods. The types of this algorithm are twofold:

1. The first type uses depths measured against geocentric vectors (GLPV).
2. The second uses depths measured against "latitudinal curves", which are at all points normal to the internal, concentric elliptical contours of the spheroid (NLPV). This latter case is to be the subject of a later discourse.

---

[18] Jefferys, MNRAS Geophyics Supplement, 3 (1935), 271-274.
[19] E Bullen, MNRAS Geophysics Supplement, 4/2 (1937), 143-157.
[20] GR Daglis &YuP Sizov, *Rapid and Concurrent Epi- & Hypocenter Location using Tabular Data Structures*, Proceedings 2ECEES Istambul, 2014.

The structure of this Chapter is briefly summarised under:

- Integral for generating trajectories on and within an ellipsoid (spheroid)
- Description and results for spheroidal localisation algorithms
- Summary of all algorithm results
- Comment on processing power needed to support spheroidal calculations
- "Stabilising" the output results
- Conclusions
- Projected further work
- Amplification of the mathematical development for the given integral (Annex I)

Table 10.1. Earthquake parameters

Number	Region	Magnitude [M(w)]	Epicentre Lat/long (degrees)	H[d] (km)
1	West of North American Coast	5.4	41.7136N/ 126.844W	9.9
2	Fiji Islands Region	4.5	17.8997S/ 178.5269W	558.9
3	Just East of Honshu	5.4	35.621N/ 140.6862E	36.0
4	South Mid-Atlantic Trench	5.2	47.186S/ 13.430W	10.0
5	Celebes Sea	4.1	3.9999N/ 123.3981E	495.1
6	Yunnan (China)	6.1	27.446N/ 103.42E	10.0
7	Santa Cruz Islands	4.7	12.2398S/ 167.135E	259.3

Integral for Generating Trajectories on and within an Ellipsoid (Spheroid) by Variational Calculus

This section gives a mathematical basis for a P2P tracer acting on and within the earth as spheroid. The coordinate transformation from geocentric latitude, $\left(-\frac{\pi}{2} \leq u : (\lambda_i) \leq \frac{\pi}{2}\right)$, and from geocentric longitude, $\left(-\pi \leq v : (\phi_i) \leq \pi\right)$, to Cartesian space frame is

$$x = r_e \cos(u)\cos(v)$$
$$y = r_e \cos(u)\sin(v)$$
$$z = r_p \sin(u)$$

Here, r_p is the polar semi-axis of the spheroid, and r_e is the corresponding equatorial semi-axis. Potential scale Factors are h_u and h_v.

Now

$$dx = \frac{\partial x}{\partial u} \cdot du + \frac{\partial x}{\partial v} \cdot dv$$

$$dy = \frac{\partial y}{\partial u} \cdot du + \frac{\partial y}{\partial v} \cdot dv$$

$$dz = \frac{\partial z}{\partial u} \cdot du + \frac{\partial z}{\partial v} \cdot dv$$

so

$$dx = -r_e \sin(u)\cos(v) \cdot du - r_e \cos(u)\sin(v) \cdot dv$$

$$dy = -r_e \sin(u)\sin(v) \cdot du + r_e \cos(u)\cos(v) \cdot dv$$

$$dz = r_p \cos(u) \cdot du$$

This leads to[21])

$$h_u^2 = r_e^2 \sin^2(u) + r_p \cos^2(v)$$

$$h_v^2 = r_e^2 \cos^2(u)$$

We also have

$$ds^2 = h_u^2 \cdot du^2 + h_v^2 \cdot dv^2$$

$$\sqrt{\left(\frac{ds}{du}\right)^2} \cdot du = \sqrt{r_e^2 \sin^2(u) + r_p^2 \cos^2(u) + r_e^2 \cos^2(u) \cdot V^2} \cdot du$$

$$= F(u,V):$$

$$V = \frac{dv}{du}.$$

By setting [22]

$$\frac{\partial F}{\partial V} = k.$$

[21]   Arfken, *"Coordinate Systems" at Mathematical Methods for Physicists*, Academic \Press,1970.

[22]   C Fox, *Introduction to the Calculus of Variations (Chapter 1)*, Dover, 1984.

we can have a required geodesic on integration

$$\frac{dF}{dV} = \frac{r_e^2 \cos^2(u) \cdot V}{\sqrt{r_e^2 \sin^2(u) + r_p^2 \cos^2(u) + r_e^2 \cos^2(u) \cdot V^2}} = k$$

This then eventually becomes

$$v\big|_{\phi_1}^{\phi_2} + \beta = \frac{k}{r_e} \int_{\lambda_1}^{\lambda_2} \frac{1}{\cos(u)} \cdot \sqrt{\frac{r_e^2 \sin^2(u) + r_p^2 \cos^2(u)}{r_e^2 \cos^2(u) - k^2}} \cdot du$$

The constant k can be found by numerical means (for example, by looking for a crossing on the abscissa and then performing a linear interpolation).

Finally, the process

$$v_i = \frac{k}{r_e} \cdot \int_{\lambda_1}^{\lambda_i} \frac{1}{\cos(u)} \cdot \sqrt{\frac{r_e^2 \sin^2(u) + r_p^2 \cos^2(u)}{r_e^2 \cos^2(u) - k^2}} \cdot du + \phi_1$$

can be evaluated numerically to give the trajectory of the geodesic, $(u,v)_i$, on the surface of the spheroid, between $(\lambda,\phi)_1$ and $(\lambda,\phi)_2$ as an explicit function of u.

Now, let's consider the construction of three-space geodesics (in other words, not lying on the spheroid surface corresponding to the abiove parameter pair. If (r_e, r_p), or the eccentricity given r_e, is given, and we have a depth point of origin (originating from a position on some geocentric position vector), then a set of geocentric position vectors may be laid down, for a set of chosen points in the surface trajectory, between the same surface start and end points as those given above. Such a family of geocentric position vectors may be allowed to construct a point-to-point ray-path arising from within the body of the Earth spheroid.

Spheroidal Scan Processing Results

There are two groups of these position vectors mentioned above. One is formed from geocentric position vectors (GLPV). The other is formed from elliptical or latitudinal position vectors (figure 10.1, NLPV). The following paragraphs give a short sketch of the main features and sequence of the spheroidal P2P process.

Each of these techniques can use a specific method to accelerate the production of three dimensional non-planar ray traces, which span its initial depth point (below the Epicenter) and its current global set of station positions

Further, each variant uses a system involving the intersection of a cone and a three-space line (a single GLPV vector or a scanned line segment for an NLPV vector) to establish the next point in the P2P ray trace. For an example of a curving NLPV, see figure 10.1.

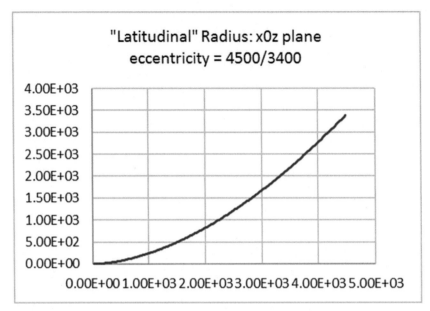

Figure 10.1. Example of an NLPV (normal latitudinal position vector)

Given an origin depth point and a take-off angle for the ray, then the following actions may take place:

1. Find the intersection point, $\underline{P}_i$, of the ray with the next position vector.
2. Calculate the next angle of incidence/refraction pair at this point, having found:
    *   the equation of the co-spheroidal ellipse at this point
    *   the bracketing isovelocity co-spheroidal or concentric contours
    *   from these contours, the velocity at the point $\underline{P}_i$, using the normals to the isovelocity contours
3. Using the information at 2, move on the pair of object position vectors.
4. Continue the process 1–3, until the set of position vectors is exhausted and note if the ray has not reached the surface of the object spheroid at the end of the given geodesic *or* the ray reaches the surface before this *or* the ray reaches the surface close enough to the end point of the prescribed geodesic.
5. Adjust the take-off angle appropriately at 1, until there is a satisfactory hit on the end point of the surface geodesic for that particular depth of origin.

At present, tests of P2P using the geocentric option show some jitter in the curves formed from the variances of the residuals of match, and other indicators (q,v. below), at each depth point in the scan. This jitter is considered as noise, and an attempt to filter it using least squares fit of order 2 and order 3 polynomials is made. These fits are differentiated to find the minima at which the target hypocentral depth, H_a, may exist.

So far, four primary sources of noise are thought to exist:

*   The calculation of the surface geodesic path – This depends on the evaluation, under varying conditions, of an integral to generate its points. The accuracy of this evaluation must be assured.
*   The calculation of the "curve constant", *k*, (see above) – This also depends on the accurate evaluation of the integral given above, to produce values for a scan to determine the point at which the required value of *k* gives a zero.

- The three-dimensional aspects of the P2P ray-trace itself – Currently, radial earth velocity models are employed, together with the usual provisos regarding the volume enclosed by each contour surface.
- The fact that the samples of the sets from the data timings and the P2P timings originating at each depth point have dissimilar cardinality.

It is hoped that the influence of such sources may be diminished after some further work.

Before this, however, we consider earthquake 5 (E05) (table 10.1) as it is used to provide general examples of indicator output. See figures 10.2 to 10.3. GLPV technique is used.

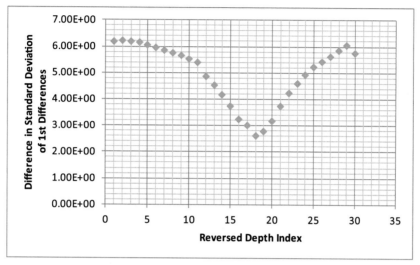

Figure 10.2. E05 – GLPV indicator type 2 (see Y-axis annotation)

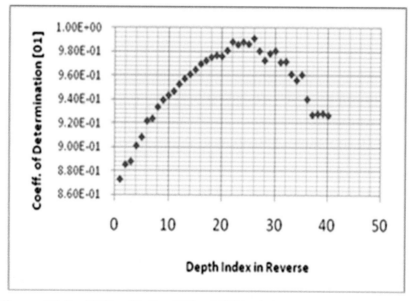

Figure 10.3a. E05 – Unity-SS(res)/SS(tot) (see Y-axis annotation)

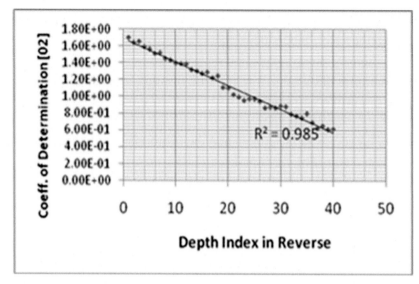

Figure 10.3b. E05 – SS(reg)/SS(tot) (see Y-axis annotation)

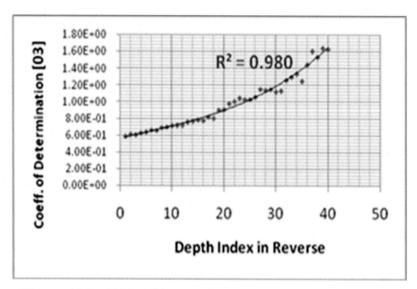

Figure 10.3c. E05 – SS(tot)/SS(reg) (see Y-axis annotation)

In figures 10.3a and 10.3b, SS(tot), SS(reg), and SS(res) are the sums of squares relating to total variance, regression, and residuals.

That is

$$SS_{[tot]} = \sum (t_i - \bar{t})^2$$

$$SS_{[reg]} = \sum (p_i - \bar{t})^2$$

$$SS_{[res]} = \sum (t_i - p_i)^2$$

where

t_i form the set of given timings, and
p_i form the set of P2P timings.

However, in figure 10.3c

$$SS_{[tot]} = \sum (p_i - \bar{p})^2$$

$$SS_{[reg]} = \sum (t_i - \bar{p})^2$$

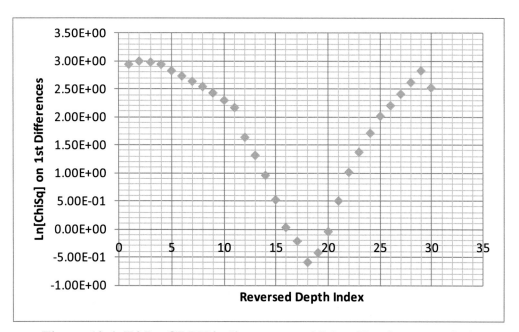

Figure 10.4. E05 – GLPV indicator type 05 (see Y-axis annotation)

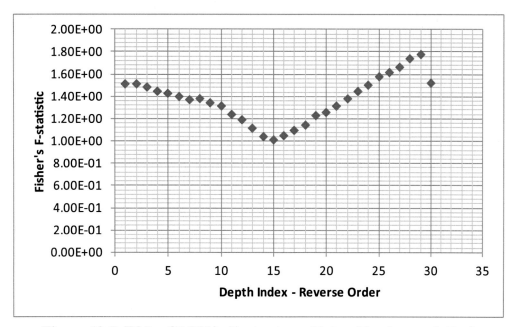

Figure 10.5. E05 – GLPV indicator type 03 (see Y-axis annotation)

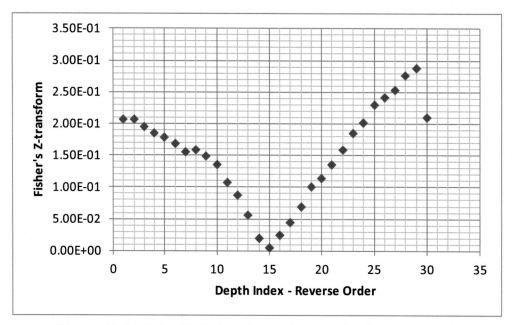

Figure 10.6. E05 – GLPV indicator 04 (see Y-axis annotation)

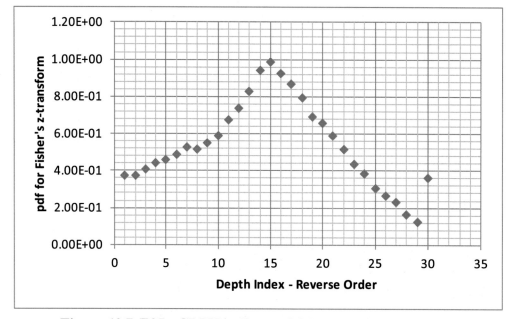

Figure 10.7. E05 –GLPV indicator 04 (see Y-axis annotation)

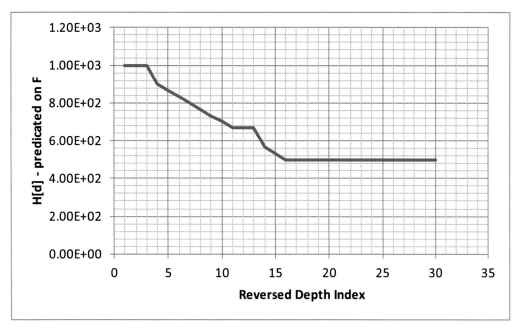

Figure 10.8. E05 – GLPV hypocentre evolution

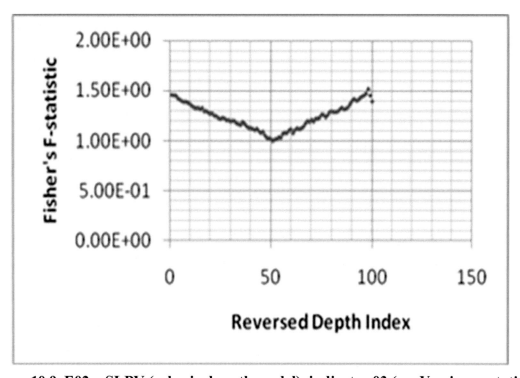

Figure 10.9. E02 – SLPV (spherical earth model), indicator 03 (see Y-axis annotation)

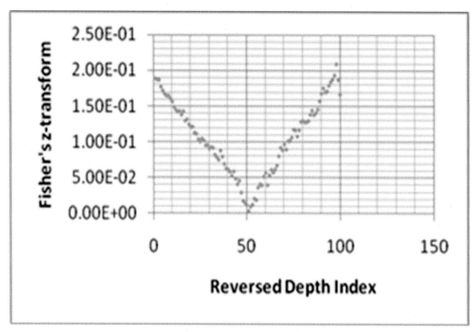

Figure 10.10. E02 – SLPV (spherical earth model) indicator 04 (see Y-axis annotation)

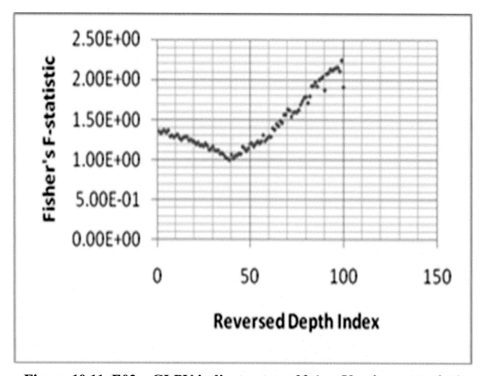

Figure 10.11. E02 – GLPV indicator type 03 (see Y-axis annotation)

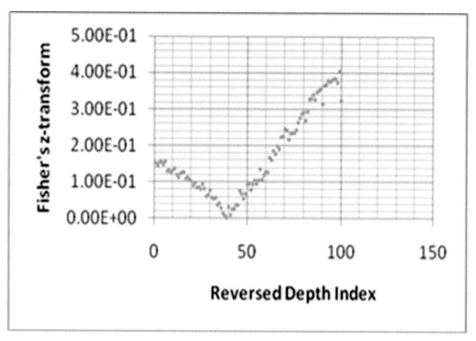

Figure 10.12. E02 – GLPV indicator type 04 (see Y-axis annotation)

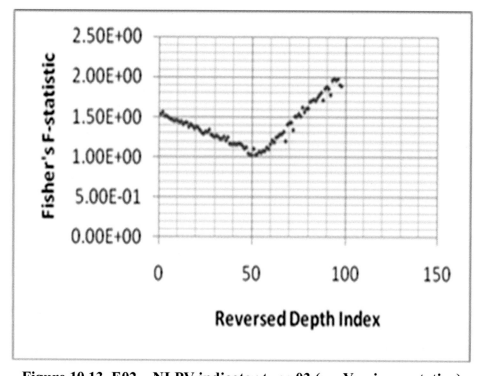

Figure 10.13. E02 – NLPV indicator type 03 (see Y-axis annotation)

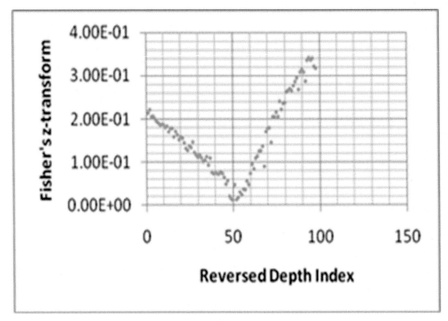

Figure 10.14. E02 – NLPV indicator type 04 (see Y-axis annotation)

Summary of Aspects of Output Results

There are two groups of results presented:

1. Figures 10.2 to 10.8
2. Figures 10.9 to 10.14

Group 1 concerns earthquake 5 (E05) (H[d] = 495 km) and uses two differing types of indicators—the centralised time differences (giving H[d] = 400km) and the Fisher statistics (giving H[d] = 500 km).

There is a significant difference between these indicators' results, both under the GLPV regime. However, the granularity of the scan is coarse. So this test might be rerun with higher granularity.

In *Group 2*, all regimes are tested on earthquake 2 (E02) (H[d] = 559 km) using a granularity of 100 depth points and Fisher statistics. Both SLPV and NLPV methods give results less than 559 km. The GLPV regime gives greater than this.

Some Attempts at Stabilising the Jitter

It will be noted that there is some degree of jitter in the trajectories of these indicators. They are intended at this point to show (by attaining, at minimum, the sought-after value for H_d) the focal depth of the event.

It was thought that a means of stabilising the indicator output in some way would be a good thing. Thus, the two least-squares fits (polynomials of orders 2 and 3) which have been used to filter this noise, might not be needed to pick the focal depth after the scan.

As mentioned above, one of the causes of the jitter would appear to be the two different sample sizes of the sets of the timings that would be used to form the indicator(s). The effect of these different sample sizes could be compensated for by choosing a statistic that could act as a stabilised indicator in this regard. Such statistics are Fisher's *F* and Fisher's *Z* – transform.

In the following, we will demonstrate some properties of four indicators. Each indicator detects the required H_d during a scan by demonstrating a maximum or minimum to which the required focal depth is inferred to correspond.

The first indicator, however, is not a Fisher statistic. Rather, it is the rms of the differences, taken at each depth point, between the centralised P2P timings and the centralised data timings from the event itself.

The second indicator is a statistic formed from the difference between the standard deviation/ variances of the P2P timings and that of the event arrival times.

The third is the Fisher *F* – statistic formed from the two sets of timings:

$$F = \frac{n_1 (n_2 - 1)\sigma_1^2}{n_2 (n_1 - 1)\sigma_2^2}$$

σ_1 is chosen always to be greater than σ_2.

The fourth indicator is chosen to be the Fisher *z* – transform:

$$z = \frac{1}{2}\ln\left\{\frac{n_1 (n_2 - 1)\sigma_1^2}{n_2 (n_1 - 1)\sigma_2^2}\right\}$$

A fifth indicator group is formed from the Chi-squared statistic used in conjunction with **basally** centered sets of the P2P and given timings as opposed to their **centrally** centered sets.

Taking earthquake 2 (E02) into consideration, we will show examples of each of these indicators acting in a hypocentre scan on a spherical and then a spheroidal model.

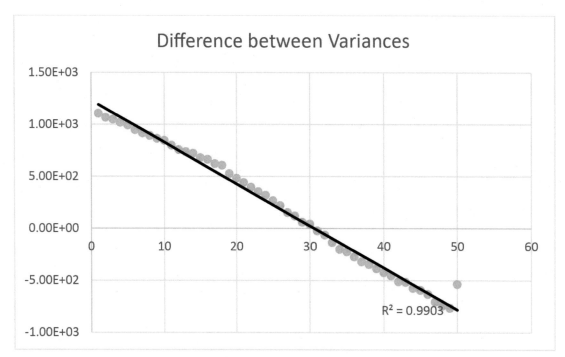

Figure 10.15: E02, SLPV

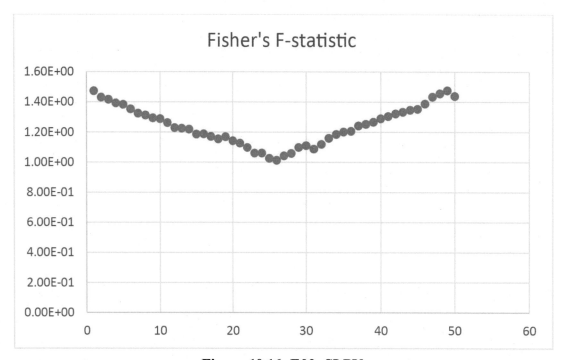

Figure 10.16: E02, SLPV

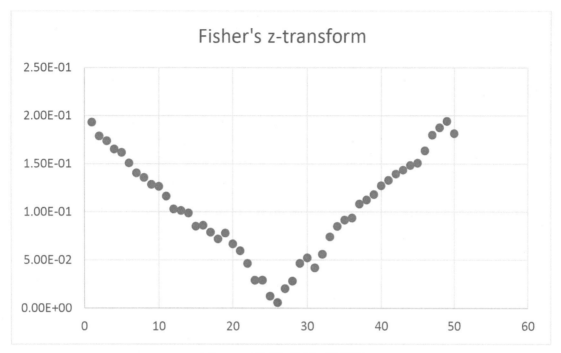

Figure 10.17: E02, SLPV

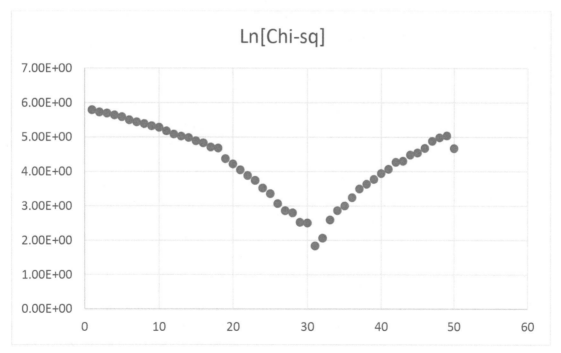

Figure 10.18: E02, SLPV

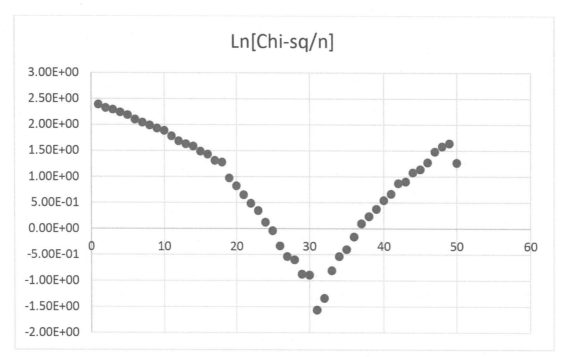

Figure 10.19: E02, SLPV

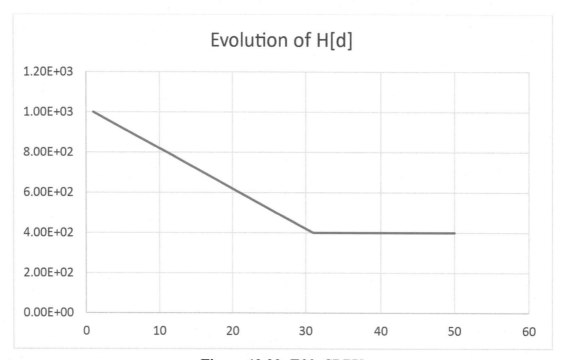

Figure 10.20: E02, SLPV

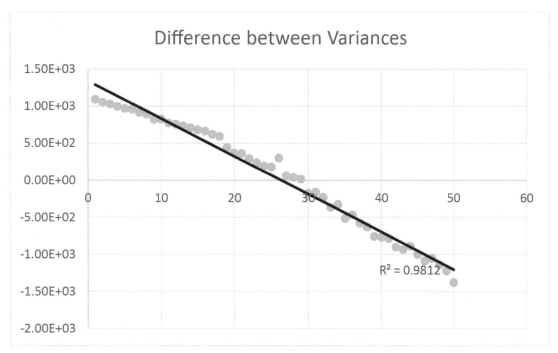

Figure 10.21: E02, GLPV (Spheroid}

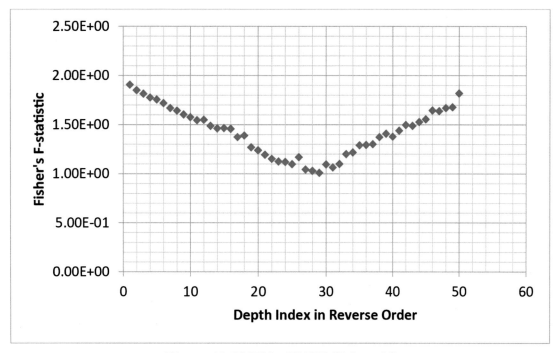

Figure 10.22 E02, GLPV (Spheroid)

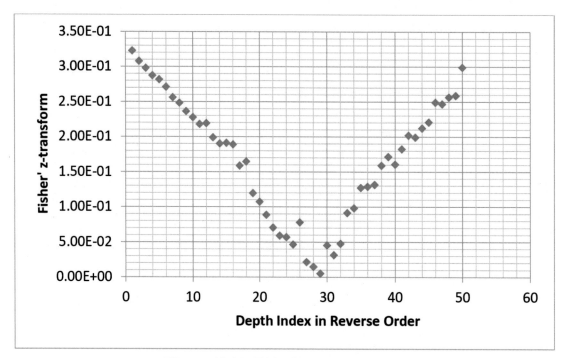

Figure 10.23: E02, GLPV (Spheroid)

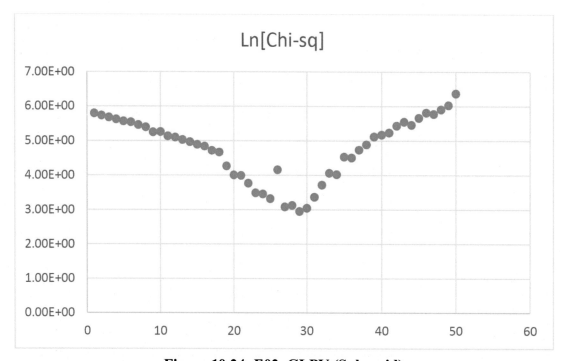

Figure 10.24: E02, GLPV (Spheroid)

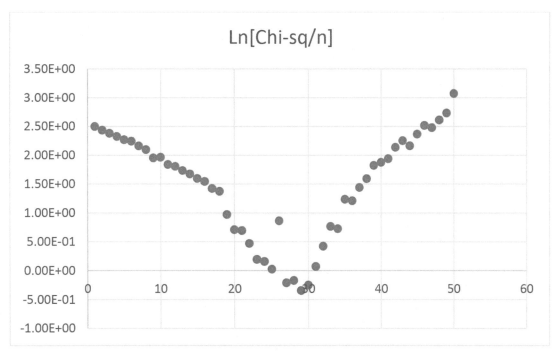

Figure 10.25: E02, GLPV (Spheroid)

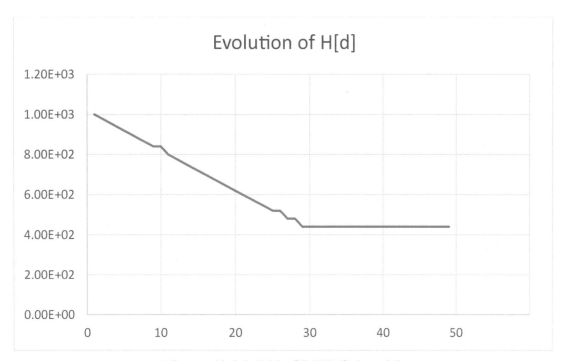

Figure 10.26: E02, GLPV (Spheroid)

The Processing Power needed for True Feasibility

The rate of processing of this hypocentral spheroidal scan is governed by several variable parameters:

1. The number of depth points at which to form a value for the rms of residuals of comparison (match)
2. The number of stations being considered (for which set there is currently data) – In the code implemented here (QlocRay00Duff06DXpress.cpp; Ellipse03Xpress.h), allowance is made for forty-five stations.
3. The number of points that are to delineate surface geodesics from epicentre (assumed known to a good approximation) to each station in the considered set
4. The number of divisions in the initial take-off angle in the primitive P2P travel time scan currently adopted
5. The convergence criterion for the integrals in the process for calculating the geodesic constant, k (see the second section of this Chapter, above)
6. The convergence criterion for each of the points in the delineation of each surface geodesic.

Let's turn next to guidelines for the setting of each of these parameters. The parameters:

1. Should generally be set to 300 to 1,000 (and the scan should start from 1,000 kilometres depth if nothing is known about the depth of the hypocentral focus)
2. Should either be a static figure (in other words, the stations are a selected starting set for which all P-wave and S-wave arrivals have been gathered) or a dynamically increasing figure (as in the "reduced" simulation test method mentioned above)
3. Can be sensibly set from 250 to 750
4. Can sensibly be set from 3,500 to 6,550
5. Should not be greater than 1.0e-2
6. Should not greater than 1.0e-3

Using an Intel i7 six-core processor in "single-thread" mode, runs with the parameters in this general region took eighteen to twenty hours to scan to completion. However, applying the compiler resource of Intel Parallel Studio XE and adapting some of the hotspots (bottlenecks) appropriately, this was reduced to one to two hours of processing. These and similar instances underwent speed factor increases of thirteen to twenty-four times.

Although these speeds may be useful in offline mode, they are not nearly enough to allow this spheroidal scan to be employed in real time or near real time.

In conclusion, a higher degree of parallelism and vectorisation would have to be found for these aspects to be both convenient and feasible.

Conclusions

It has been shown (see result *Group 1* and result *Group 2* in Results Section, above) that algorithms that directl use of Earth's spheroidal geometry can be implemented in scans and will produce results compatible with that which is known to be the epicentral and hypocentral parameters of those earthquakes considered (taken from table 10.1). The slight differences between the values calculated and the Wilber

3 database values are observed qualitatively here. This implies that a more stringent analysis could be conducted numerically to quantify the performance of spheroidally modified algorithms.

Intentions for Further Work

Although a considerable database has been accumulated covering localisations for each of the earthquakes listed in table 10.1, time and space have not allowed a full expression (quantitatively and qualitatively) of what it holds. Only a demonstration of feasibility has been possible.

Further work would lead to the inception and expansion of a numerical analysis for the spheroidally modified localisation algorithms on each of the earthquakes listed in table 10.1. This would aim to show what changes are made, if any, between the tabular added corrections for ellipticity [23, 24] and the direct use of spheroidal geometry in event localisation.

---

[23] Jefferys, MNRAS Geophyics Supplement, 3 (1935), 271-274.
[24] Bullen E, MNRAS Geophysics Supplement, 4/2 (1937), 143-157.

Acknowledgements

Acknowledgement is made to Incorporated Research Institutions for Seismology (IRIS) for providing the earthquake database handled by Wilber 3, without which this analysis would not have been possible.

Annex A
Description of System Entities

Entities Contained in This Description

1. **Data/Information**
   a. Text files
   b. Console or screen input
   c. Graphical output
       (a) Depiction of seismograms
       (b) Depiction of energy onset within individual seismograms

2. **Programs/System Nodes**
   These entities, when activated, cause the general flow of data within the system. They are supported by the following groups of functions:
   - **EarthQuakeLocationHeader.h**
   - **EikonalScanningTools.h**
   - **EpicentreLocationHessian.h**
   - **FFTandCorrelationD.h**
   - **FFTandCorrelationF.h**
   - **HypocentreConsolidationCC.h**
   - **HypocentreConsolidationLL.h**
   - **HypocentreScanningTimeBased.h**
   - **HypocentreScanningTimeBasedD.h**
   - **HypocentreScanningTimeBasedF.h**
   - **HypocentreScanningTransBased.h**
   - **HypocentreScanningTools.h**
   - **IntegrationTools.h**
   - **QuakeLocationGaussNewton.h**
   - **QuakeLocationHeader00.h**

3. **External Systems**
   a. Wilber 2 or 3 (IRIS)
   b. Excel 2005 (Microsoft)
   IRIS refers to the very excellent Incorporated Research Institutions for Seismology. The system Wilber 2 (and its beta version Wilber 3) is a means whereby users from within IRIS can extract seismological data for their own research purposes.

   The Microsoft subsystem Excel 2005 is here used, under the Windows XP operating system to generate graphical output. This will contain cues, which will ultimately form input to the system from the screen/console and so is crucial to the attempts by the system to locate epicentral coordinates (although, this pathway is, in its current form, slow).

Description of the Major Entities

Data/Information

These are depicted in red in the system charts provided in diagrams B01, B02, and B03.

GlobalData. This file is initialised during the input of seismogram data and serves to supply the following parameter set to each node in the system, where required. The values are:
- Number of stations forming the current batch
- Number of seismograms originating from each station
- Total number of seismograms
- Number of sample points in each seismogram
- Sample interval (seconds)

00. Contains the raw seismograms, as initially extracted from IRIS, but de-multiplexed into one file. Each sample row is annotated with a time index. This de-multiplexing involves a unifying time-fusion process for multiple seismograms input from each station.

01. Contains the output necessary for constructing energy-onset diagrams for the set of seismograms in **00**. Each of the blocks of graph data is keyed to its station/seismogram.

02. Contains augmented positional information for each station corresponding to the set of seismograms in **00**. This information is latitude/longitude, elevation, depth below ground, earth radius 1, and earth radius 2. The time of the inception of the seismogram is also recorded as year, day, hour, minute, second, and millisecond. Also it contains the event epicentre coordinates derived from the **SAC(A)** header input.

03Saved. A condensation of the **02** data. However, it simply contains the (x, y, z) coordinates for each station, using the earth space frame and the seismograms in **00**, together with their initial "Time - zeros", relative to an arbitrarily given point in time.

04. Contains a catalogue of the discovered onsets (as found in **01**) together with time index and their timings under the scheme described in **02** and **03**. Each entry is also keyed to station/seismogram.

At separate stages to these above usages, the files **00, 01, 02, 03, 04** are used by the tripartite table-generating routines incorporated in

TimeBased01.cpp

TimeBased01CNV.cpp

The files **Combinations00AA** to **00QQ** are annotated holding files for previously generated tables. They each singularly hold the contents of **00** (timings), **01** (take-off angles), **02** (calibration), **03** (error indications), and **04** (failure reporting). The transportation of these large tabular matrices is performed by three nodes:

- **MXTST 00.cpp** (unique source [single volume] to unique target [single volume])
- **MSTST 00.cpp** (many sources [single volumes] to one [multivolume] target)
- **MQTST 00.cpp** (one [multi volume] to many [single volumes])

(Note: In the current state of the system, **01** and **04** may be bypassed, since they are serviced by nodes that assume only one seismogram per station. Subsequently, they have been replaced by **06** and **07**, which perform the same service, respectively. See below.)

05. An intermediate file for passing initial Locations made by the scanning processes in

DynamicSphericalDeterminationAvto 03, 04

to

StaticSphericalDeterminationAccelerated,
StaticSphericalDeterminationBypass

in order to see if statistical refinements can be made.

06. As **01**, contains the information from which the "δ-functions" in the energy-onset diagrams can be graphically generated. This file is served by nodes that have combined multiple seismograms at each station into a single "time-fused" data entity (that is, combinations of N, E, and/or Z seismograms from each station).

07. As for **04**, only as **06**, contains information from the fusion of combinations of traces at each station.

08. Currently void.

09. Currently void.

10. Currently void.

11. Monolithic file containing a "gather" on all active elements of the set of **Phasordd** files output by **FTTST 02**. The gather or "fan in" is performed by **PHTST 00**.

12. Monolithic file resulting from a "gather" or "fan in" on the output files of **IGTST 04** or, more importantly, **05, 06,** or **07** by the data handling routine **ISTST 00**. These output files are collectively the **InSeisdd** file set and are the integrated versions of the Fourier transforms contained in the **Phasordd** set.

13. File containing output from **PWTST 06**, namely the reduced set of detected onsets. 13 is used by all programs that wish to determine onsets by the ordinal number of their appearance on the reduced energy-onset diagrams, now held in **06**, and also importantly in the set of files **OutSeisdd**, where the detected onset points are shown superimposed on the individual seismograms (see Annex B). This file is the source of onset timings and indices for the node **AUTST 01**.

Phasordd. A cluster of files that contain individually the Fourier transform output from **FTTST 02** (which acts also as input to the integration routines **IGTST 04** and **IGTST 05, 06, 07**) for each seismogram, be it accelerogram or velocigram.

InSeisdd. Contain the individual integrated seismograms provided by the pair (**FTTST 02, IGTST 05, 06, 07**) and originating at **00**.

FileLibraryIN00. A set of filenames held for the purposes of making the file handling in **IGTST 07** dynamic, as does **FileLibraryOUT00**.

Seismograms00. A backup work file for the de-multiplexed versions of the energy-onset detection processes **PWTST 05, 06, 07, 08** in whatever form. Seismograms00 contains, at the end of the process, the set of seismograms, transformed to oscillate about their means.

SeismogramsXTR. Contains a structure formed from slimmed-down versions of all traces in **Seismograms00**, in other words, traces in which every nth (specifiable) sample-row is retained.

OutSeisdd. Contains the data that, if processed by **Excel**, will provide diagrams of the energy onsets superimposed upon their individual seismograms. (See Annex B.)

AutoData01. Concerned with carrying an approximate epicentral location between the nodes **DLTST 01, DPTST 01, DQTST 01,** and **AUTST 01**. The file aslo contains residual information about this Location, ε, but this is not yet used by the subsequent nodes, although it is displayed to the user by the node **DLTST 01**.

AutoData00. An intermediary between the nodes **AUTST 01, AUTST 02, AUTST 03** or **AUTST 04** and the nodes

DynamicSphericalDeterminationAvto 03, 04

It consists of sets of actual timings, relative to the initial point of the lead station (with time-fused seismograms) and arranged in lists for each station, together with the set of local indices (keys) to the timings found in each "record" corresponding to the seismograms to be found in **07**.

Velocity00. Used as part of the system final output from the node

StaticSphericalDeterminationAccelerated.

It is a detailed audit trail for the iterative calculations made by this node.

Velocity01. Contains hard copy of the final summary results of the system. These form a subset within **Velocity00**.

Velocity02. Used to pass the single best conclusion for the epicentral Cartesian coordinates (together with its corresponding value for the radial discrepancy as a percentage of a mean earth radius) on to the terminating set of routines in the system

EarthQuakeLocation07, 08 or 09
ETTST 01, 08

This set of routines is briefly described below. Sometimes the output from

StaticSphericalDeterminationAccelerated

may be bypassed by using the routine

StaticSphericalDeterminationBypass

which moves a single selected output of epicentre coordinates from the file **05** (q.v.) without further refinement or processing to **Velocity02**.

VelocityModel00, 01. Contain a model (for example, the preliminary reference earth model [PREM] by Djeiwonsky et al.) or list of the various velocities (P and S wave) and densities to be encountered within the sphere of the earth. They arc input to the ray tracing facilities within the above node.

VelocityProfile00. Contains the complete version of ak135 as published by B. L. N. Kennett et al. in 1995.

RayPath00. Contains lists of the take-off angles for the rays traced to each of the selected seismic stations involved with the location of the hypocentre, which is considered to be immediately below the epicentre. the calculation of the hypocentral depth involves a scan. each point in the scan generates a set of rays to the set of stations.

RayPath01. Contains the main results of the scan, being a curve made up of the rms residuals generated by the process for each point on the ascending scan. This will enable a graphical indication of the locality of the hypocentre to be given. In addition, this file contains a more accurate result read off the curve as it was being formed by the routine. Thus, from this file and from **Velocity02**, the position of the given seismic event can be inferred.

EpicentrePolar02. Contains the value of the latitude and longitude for the object seismic event output by any of **DLTST 01**, **DPTST 01**, or **DQTST 01**. This may be input by either of the routines **ETTST 01** or **08**, which perform the interpolative table-driven scan for the hypocentre focal depth.

EpicentrePolar03. Contains a given value of the latitude and longitude for the object seismic event (possible from IRIS's Wilber 2 or 3). This information is derived by **RDTST 06, 07** from the input **SAC(A)** header and passed to this file by **RDTST 02** from **02** (q.v.). This may be input by the routine **ETTST 01** or **08**, which perform the interpolative table-driven scan for the hypocentre focal depth as well as by **EarthQuakeLocation 07, 08**, or **09**.

HypoCentreDataRR. Generated by **RQTST 00** from the contents of the permanently saved data in **03Saved**. It is used as input to **HCTST 00** when this program is serving the pathway that leads to the programs **SQTST 00** or **02**.

HypoCentreData. An output file for **RDTST 02**, along with **03Saved**. It is also input data for **ETTST 01** or **08**. Further, it is used as input for **SQTST 00** or **02**, having been set up by **HCTST 00**, which uses **HypoCentreDataRR** as input.

HypoCentreResult. The result of the hypocentre determination by the tabular scan performed by **ETTST 01** or **08**.

HypoCentreResultSQ. Holds the result for the "fast track" end routines **SQTST 00** and **02**. It contains the located coordinates of the epicentre and hypocentre, together with the take-off angles to each active station.

PWaveTimeIndices. A a repository for picks (for P arrivals) that is used, together with **03Saved**, to generate the **HypoCentreData** entity, given above. The program node that does this is **HCTST 00**. This file is filled by manual input via the "pick" routines, **IPTST 00, 01** from scrutiny of the **OutSeisdd** files.

SWaveTimeIndices. A repository for picks (for S arrivals) that is used, together with **03Saved**, to generate the **HypoCentreData** entity, given above. The program node that does this is **HCTST 00**. This file is filled by manual input via the "pick" routines, **IPTST 00** and **01** from scrutiny of the **OutSeisdd** files.

LWaveTimeIndices. A repository for picks (for L arrivals) that is used, together with **03Saved**, to generate the **HypoCentreData** entity, given above. The program node that does this is **HCTST 00**. This file is filled by manual input via the "pick" routines, **IPTST 00, 01** from scrutiny of the **OutSeisdd** files.

RWaveTimeIndices. A repository for picks (for R arrivals) that is used, together with **03Saved**, to generate the **HypoCentreData** entity, given above. The program node that does this is **HCTST 00**. This file is filled by manual input via the "pick" routines, **IPTST 00, 01** from scrutiny of the **OutSeisdd** files.

StationPolarTransform. Output from **ETTST 01** or **08** containing the results of the geographical coordinate transformation of the object seismic stations and the results (in graphical terms) of the tabular scan.

PSArrivalTabulation00. Contains the active set of tabulated timings from **00** or any of **Combinations00AA** to **00QQ** for immediate input into **ETTST 01** or **08** and also **SQTST 02**.

TakeOffAngleGrid. Contains the active set of tabulated take-off angles from **01** or any of **Combinations00AA** to **00QQ** for the immediate use of **ETTST 01** or **08** and also to **SQTST 02**.

In the case of both of these above files, the tabular data matrices that they are to contain are transported by either the node **MXTST 00** or the node **MSTST 00** (single and multiple target files, respectively).

StationLatLong. Output by **TimeBased01** when generating the tripartite tabular structure necessary for the scan. It contains the vectors that calibrate the latitudinal displacements and the depths used to form these tables.

Programs/System Nodes

These are represented in **blue** and contained between **black** square braces on the accompanying system diagrams, which are attached to this annex.

RDTST 04 to 06 (SAC), 07 (SACA). Collectively, the entry node to the system. It collects and places a disparate set of seismograms. First the combinations, if any, of seismograms (N, E, Z, and so on) at each station are "time-fused". Then the collected files are placed in **00**, while the rudimentary header information (including the given epicentre latitude and longitude for the object event) is placed in **02**. These routines are activated from one seismogram to three seismograms per station respectively.

RDTST 02. A program that acts to crystallise the information in **02** and passes out earth space frame coordinates (international ellipsoid) at each station, to **03Saved** and **HypoCentreData**, together with the t_0 (T-zero) for each station's seismograms relative to some arbitrary datum. Also, the given epicentre coordinates, precalculated for this event in **Wilber 2/3**, are passed to **EpicentrePolar03**.

FTTST 02. Performs FFT on the Monolithic input file **00**. Output of the transform goes to the elements of the set **Phasordd**. Each of these elements corresponds to the output of a single seismogram from **00**.

IGTST 04. Performs the integration from the original seismogram to either velocity or displacement. It uses a very slow IDFT and has been discarded, more or less, except for checking the results of the "inverse FFT" in **IGTST 05, 06, 07**, below. The output of both **IGTST 04** and **IGTST 05, 06, 07** is collected as individual seismograms in the set **InSeisdd**.

IGTST 05, 06, 07. As for **IGTST 04**, except that it uses FFT for IDFT (inverse discrete Fourier transform) and is extremely rapid. The mechanics of its workings are discussed in reference (5). **IGTST 07** uses the dynamic file handling enabled by the data held in **FileLibraryIN00, OUT00**.

ISTST 00. Performs a "gather" or "fan in" on the output files of **IGTST 04** and **IGTST 05, 06, 07** into **12** as a monolithic file, containing all the seismograms currently under consideration.

PWTST 05. The node at which all gathered seismograms, **12**, are transformed to mean, with the help of **Seismograms00** as a backing store. Output information is sent to **01**, enabling the plotting of the energy-onset diagrams. The pooled onset timings for each station, from the combination of seismograms that may be present, are passed to **04**. Each "record" or "module" within the structure of **04** is of variable length and is keyed with station number; timing ordinal number and sample number.

PWTST 06. A routine that performs a refinement on the onsets provided by **PWTST 05**. This action serves to clarify, if necessary, the main onsets from P-, S-, L-, and Rayleigh energies and to reduce and so clarify, if necessary, the number of onset indications found by **PWTST 05**. Reduced onset information is transmitted to diagrams (onset points superimposed upon their seismograms) by **OutSeisdd** files via **Excel**. An important output of this program is also **04**, which contains the reduced data from which users may identify onsets, from viewing the onset lines on the graphs generated by the **OutSeisdd** files via **Excel**.

PWTST 07. As for the foregoing **PWTST 05** but uses averaged power spectra (PSD) as a ratio to detect variations in the input energy as recorded in the seismogram

PWTST 08. As for the **PWTST 05** but has the ability to define an allowable length of the gathered input seismograms.

PWTST 09. As for the foregoing **PWTST 05** but uses a dual test structure consisting of a t-test (for means with non-homogeneous variances) followed by a process control outlier test set at a definable multiple of the current standard deviation of the long buffer.

PWTST 09a. As for the foregoing **PWTST 09** but uses a dual t-test structure consisting of a t-test proper followed by a t-test based on that for non-homogeneous variances.

PWTST10. As for the foregoing **PWTST 09** but uses a dual test structure consisting of a comparison of the natural logarithms of the two means (long and short buffers) in a t-test, as the initial test, followed by a process control outlier test.

RQTST 01 or 02. Inputs from **03Saved** on the pathway to serve the routines **SQTST 00, 01** or **SQTST 02** or **03.** These latter routines provide concurrent locations of epicentres and hypocentres in a single scan, purely using P-wave first arrivals. **RQTST 01** is responsible for providing the coordinate transform to make the action of the **SQTST 00, 01** routines tractable. This, in the forward sense, is a yaw matrix, followed by a pitch matrix, both of which are recovered in inverse form from the intermediate files **HypocentreData** and **HypocentreDataRR**.

HCTST 01 or 02. Inputs **PWaveTimeIndices** or **SWaveTimeIndices** and **03Saved** and outputs this information in a coherent table, **HypoCentreData**, for use by **ETTST 01** or **08**. Also **LWaveTimeIndices** and **RWaveTimeIndices** could be included. Apart from these standard inputs, they also input from the file **HypoCentreDataRR**. **HCTST 01** inputs the yaw and pitch matrices for the station coordinate transforms to **HypoCentreData**.

  SQTST 00, 01, 02 or **03** are important facets of this system in that they provide epicentre and hypocentre locations simultaneously in one pass of a rapid scanning process. They use as data the P-wave first arrivals processed and stored in **HypoCentreDataRR** up to and including **HypoCentreData**. Each of these routines input from **PSArrivalTabulation00** and **TakeOffAngleGrid** in order to set up their interior tables.

  The **RDTST 07; RDTST 02; RQTST 01, 02; HCTST 01, 02; SQTST 00, 01, 02 or 03** is a reduced pathway leading to a rapid event location in terms of a concurrent solution for the epicentre and hypocentre in an interpolative tabular scan by the last set of routines, **SQTST 00, 01, 02 or 03**, which are also dependent on the tripartite data structures as evinced by **00AA** to

00QQ and fed in via **00**. It also uses **StationLatLongAA** to **StationLatLongQQ**, fed in via **StationLatLong**.

In the current state, **RQTST 01** makes an estimate of the epicentre in one of two possible ways, using the most proximal stations. This estimate can be used by **SQTST 00, 01 or 02** (through the auspices of **HCTST 01**) to orientate the station coordinates and to set up the co-latitude template needed for the "type 1" indicator rms system (not **SQTST 02**). **SQTST 03** is completely free of the need for any epicentre estimate and acts on the raw station coordinates (through **RQTST 02** and **HCTST 02**). **SQTST 01** uses the "type 1" indicator rms system, while **SQTST 02 and 03** use the "type 2" indicator rms system.

SPTST 01. Contains the mechanism of **IGTST 05** as a subset and provides "bandpass" power density spectra for individual seismograms, **Phasordd**.

PHTST 00. Acts as **ISTST 00**, q.v., to "gather" the separate **Phasordd** elements into **11**.

XRTST 00. At this juncture, **XRTST 00** may be invoked to act to reduce the size of the seismograms in **Seismograms00** so that they can be plotted by the **Excel** system. The "reduced" seismograms are stored in **SeismogramsXTR**.

CPTST 01. Compares two matrices (for example, the timing tables from **01** and so on), the second relative to the first.

PRTST 01. Prepares matrices (for example, the timing tables) for display in 3-D by **Excel**.

IPTST 00, 01. Routines that provide a means of manually inputting P-, L-, S-, and R-wave "picks" of time-onset indices to the four files:

P-, SWaveTimeIndices

L-, RWaveTimeIndices

These picks are mainly made from the scrutiny of the **OutSeisdd** files and other output of the **PWTST** family. **IPTST 00** operates with separate column vectors for the P-, S-, L-, and R-wave onsets, while **IPTST 01** operates with the four vectors all in parallel, across the set of stations, so that all the fourfold information from a single seismogram can be taken in at one sweep.

DLTST 01. A routine that needs three to four or more time differences for two different surface-wave species, arriving on approximate great circle arcs, together with assumed velocities, to determine an approximating value for the epicentral coordinates of the originating event. This is used in conjunction with the program **AUTST 01** to provide, automatically, timing parameters for the dynamic and static location routines. The mathematics for this version of the location process is given at references (3), (4), and (6) in this series. As will be found for **DPTST 01**, this calculation need not exclusively be for use with L-and R waves but can be tried out on P and S waves if one assumes they have been travelling in circular arcs and ducted in some way through the lithosphere or crust.

DPTST 01. A routine that requires, as **DLTST 01**, at least three to four duplets formed from any pair of seismic wave species (in other words, between any pairing of the P-, S-, L-, and R-wave species), whose velocities (and their uncertainties) are assumed known. Unlike DLTST 00, however, the calculation assumes that the onsets forming the duplets had been created by waves traveling over a planar surface and, in the case of the P and S waves, have been ducted close to this surface. Again, as in **DLTST 01**, the first phase linearised least-squares calculation is reinforced by a Gauss–Newton "descent".

DQTST 01. A further routine involving a linearisation process. It acts in a framework similar to that of **DLTST 01** and **DPTST 01**. If a group of stations is thought to be within 4.5 degrees of distance from the epicentre of the event in question, then this routine can be applied, using time duplets. A definition of the basic linearisation process for this routine can be found in the main text of the reference (6).

The set of nodes

- **AUTST 01** (uses o/p from **DLTST 01**, **DPTST 01** and **DQTST 01**)
- **AUTST 02** (inputs surface wave onset indices from **L-** or **RWaveTimeIndices**)
- **AUTST 03** (inputs body wave onset indices from **P-** or **SWaveTimeIndices**)
- **AUTST 04** (EXACTLY duplicates **AUTST 02**)
- **AUTST 06** (inputs onset indices from P, S, L, or R waves with dynamic index stacking widths at each station)
- **DLTST 00** (uses velocity duplets on a spherical earth}
- **DPTST 00** (uses velocity duplets within a zone assumed proximal and "flat")
- **DQTST 00** (uses velocity duplets with stations projected onto a common plane) supply onset information to these epicentre fixing routines:
  - **DynamicSphericalDeterminationAvto 03, 04.cpp**
  - **StaticSphericalDeterminationAccelerated.cpp**

The latter node provides, currently, a single value for the epicentre that can be passed to **EarthQuakeLocation07, 08, 09.cpp** or **ETTST 01, 08**. These routines scan to find the hypocentral depth corresponding to this estimation for the epicentre. The former node scans using ray trace routines directly, while the latter pair of nodes performs an interpolative table-driven scan (thereby using the ray trace routines indirectly).

AUTST 01. In order to facilitate the input of onset and station parameters to the dynamic and static spherical determination routines, the programs **AUTST 02** and **AUTST 01** were provided. This latter is an initial form of this automated process, in that it follows on from the discovery of the epicentral coordinates by **DLTST 01**, **DPTST 01**, and **DQTST 01**. It proceeds as follows:

1. Input basic steering parameters:
   - Degree of seismogram grouping per station
   - Sampling interval for these seismograms
   - Largest size for the timing data stack per seismogram
   - Initial point of onset in lead seismogram for wave species, defined by index or sample number
   - "Group" velocity for the wave species
   - "Spread" of velocities for this "group" value
2. The subsequent onset-points within the remaining set of seismograms are thus calculated knowing:
   - The epicentre of the event to be monitored
   - The positions on the reference sphere of the set of seismograph stations
   - The assumed lower and upper limits on the velocities of the given wave-species

3. The upper and lower indices for the timings on the samples, which correspond and possibly bracket the onsets mentioned above are also calculated. These indices are derived from the file **13**, which contains a list of all energy onsets discovered by the program series **PWTST dd** (in particular **PWTST 05**) for each seismogram (or combined seismograms) in the overall set.

4. These results are then displayed to the user for selection purposes, and so, if required, a subset of the total set of seismograms can be defined.

5. The subset selection can be repeated until satisfied.

6. The parameters extracted by this process consist of sets of actual timings, relative to the initial point of the lead station, or seismogram, and arranged in lists for each station together with the set of local indices (keys) to the timings found in each "record" for the seismograms to be found in the file **13**.

The further routines **AUTST 03** and **AUTST 04**, unlike **01**, do not use any initial estimate of the epicentre to bracket the "sheaves" of timings to be used in the scanning processes that attempt to define the epicentre more accurately. Instead, they use the output of **PWTST 05** and/or **06** indirectly, via the action of **IPTST 00, 01**, to set up centre timings to bracket sheaves at each active station. **AUTST 03** uses nearest neighbour timings and is a fine-tuning routine, while **AUTST 04** uses neighbouring onset timings that have already been picked out by **PWTST 05** and/or **06**. Therefore, this latter gives rise to a scan based on a much coarser scope. (However, results from such a scan can be passed to **AUTST 03** to provide a more refined result set.).

AUTST 06. Accepts all four types of onset timing indices (for P, S, L, or R waves) and allows the input of values of the width for timing index stacking to be applied separately at each station. These widths should reflect an uncertainty in each timing index.

DynamicSphericalDeterminationAvto 03, 04. Routines that perform one form of combinatorial scan using the sets of timings provided at each station by the program **AUTST 01**. The results of successful scans are collected for each lead timing employed. The epicentre that corresponds to the most coherent result, for this lead timing, bearing in mind the propagation velocity and time of event are being fixed together with the epicentral latitude and longitude, is recorded into **05**. For each epicentral candidate so recorded, the local ordinal numbers that define its constituent timings are also recorded. On view to the user is each fix made—together with earth radial discrepancy, time to origin, internal consistency, rms residual, and the number of combinations scanned for this trenche of input. Each trenche is headed by a particular lead time, (taken again from the lead station). The mathematics used for the fix of the epicentre in these programs is given at (6).

StaticSphericalDeterminationAccelerated. This routine attempts to refine the results passed to it in the file **05**, chiefly by re-weighting the residuals as a function of individual velocities calculated from the source to each station—in other words, not necessarily assuming ideal and isotropic propagation conditions. The final output from this routine goes four ways:

1. The **Screen Monitor** for summary results, as they happen
2. **Velocity00** for a detailed audit trail of the calculations made

3. The file **Velocity01** for hard copy of the summary results
4. **Velocity02**, which communicates the final result both to the observer and to the next node in the sequence (see below)

StaticSphericalDeterminationBypass. This routine, having allowed the selection of a unique result to take place from those in the file **05**, passes this result, so selected by it, and without further processing, to the file **Velocity02**. In this case, the other **Velocity00** and **01** files are bypassed and are null.

EarthQuakeLocation07, 08 and **09**. Collectively, form the final node in the sequence, which currently begins with the routines **RDTST dd**. This node accepts the output of the preceding node

StaticSphericalDeterminationAccelerated

which has radiated into

Velocity00
Velocity01
Velocity02

The main input consists of the Cartesian coordinates of the epicentre placed by and resulting from this latter node. These are to be found in **Velocity02**. As stated above, other automatically transmitted input is the PREM (or other) one-dimensional velocity model for the interior of the earth (**VelocityModel00, VelocityModel01**); the set of Cartesian coordinates (relating to the earth space frame) for each of the seismic stations in the group currently under consideration (in other words, the group of stations corresponding to the seismograms that have been imported into the system) (**03**); and the set of station-related energy onset times (**13**). These can, by virtue of the output of graphical diagrams—in which onset markers are superimposed upon the individual seismograms (after first- or second- level Integration)—be used to pick the corresponding and actual arrival times.

This routine scans vertically upwards toward the epicentral point from a specified depth and using a specified step length for the ascent. At each part of this ascending trajectory, a ray is generated (P-wave rays in the case of **07**, both P-waves and S-waves rays in concert in the case of **08**, and S-wave rays only in the case of **09**) to reach each of the selected seismic stations. This point-to-point ray tracing takes place through the spherically curved strata of crust, mantle, and core, using the PREM scheme to supply the velocity information. This process ultimately defines the travel time for each ray species.

Like the initially recorded and, therefore, known arrivals that are presented as offsets grouped around their centred mean, the above set of travel times is centred about its own mean value, and offsets are obtained. These are then compared with the observed set of offsets. These comparisons provide a root mean square (rms) residual for a given point. These rms residuals are tracked, as they are formed, upwards according to the tracking parameters entered. The depth corresponding to that rms value, which is the infimum of the set, is taken to be an indication of the position of the hypocentre.

Figures **B01**, **B02**, and **B03**, below, are apt to demonstrate this process. Output files are **RayPath00** and **RayPath01**, as described above.

ETTST 01. This node provides scansion from tables by interpolation (linear, cubic, or Lagrange) as described in the foregoing. It inputs directly from the tables of timings held in **PSArrivalTabulation** and from the tables of take-off angles from the file **TakeOffAngleGrid**. It selects a velocity model from one of the above-mentioned Earth Velocity Model files and inputs the Seismic Station data (coordinates) from **HypoCentreData,** which it uses to scan against the given set of tables across a user-defined set of depths and latitudinal displacements by interpolation. Output of results goes to **StationPolarTransform** and **HypocentreResult**. The epicentre coordinates necessary for this operation are input by choice, either from **Velocity02** or from **EpiCentrePolar02** or **EpiCentrePolar03**.

ETTST 08. As **ETTST 01** but with a gating system for selecting a subset of the active seismic stations. Also in refining the results of the hypocentre scanning, a version of Chauvenet's principle for the rejection of outliers has been implemented in **ETTST 08** as an incremental one-by-one process. This process can recycle until either it is recommended that no further stations be dropped as outliers or the number of stations is deemed to have been reduced too far. It is entirely the prerogative of the user as to whether he or she complies with any such recommendations. The recommendations can be rejected at any point, and the process can be recycled back to its start. It may also be possible to incorporate a weighting scheme based on the outlier principle and the perceived uncertainty in the picks of the timings of the P- and/or S-wave onset energies.

Further programs **UETST 00** and **UETST** 01 can be used to qualify the hypocentral results. The single epicentre position, as used by **ETTST 00** and **ETTST 01**, is multplied into an azimuthal grid (in the former) or a rectangular grid (as part of the spherical manifold in the latter. Each node (possible epicentre) in this grid is used, after due station coordinate transform, as the basis for a scan. The epicentral node that corresponds to the smallest rms scanning residual is selected, together with its corresponding value for the hypocentre depth. Some of the workings of these programs is illustrated in Annex C.

ETTST 04 to 07. Adjust and compensate for error, in other words, inaccuracy of the point-to-point (P2P) ray tracers used to generate the scanning tables by **TimeBased01**. They are also used to preprocess and fine adjust scanning tables before their use.

(The mathematics central to all those latter routines, which deal with FFT, onset detection and location of epicentre or eventually hypocentre, is briefly taken up in chapter 3).

External Systems

These lie on the periphery of the processing system whose main components are described above.

Wilber 2 or 3 (IRIS). Serves as a source of a set of **Individual Seismic Analysis Code (SAC or SACA) ASCII** files, each of which contains a selected seismogram. These seismograms are selected relative to a chosen event in terms of radial displacement; selection of (N, E, and Z) components; and signal to noise. All these factors are considered on stations within those seismological station networks selected by the user.

Microsoft Excel 2005. A Microsoft subsystem used for the generation of graphical representations of:

1. Seismograms at various levels of integration and resulting from various bandpass windows
2. Power spectra
3. Energy-onset diagrams
4. Superimposition of energy-onsets upon seismograms

These forms of output are cues for the selection of values defining time eifferences required in the case of the nodes **DLTST 01, DPTST 01** and **DQTST 01**, for instance, and also those onsets required by the final node systems **EarthQuakeLocation07**, **08**, or **09**, and **ETTST 01 and 08**. As such, they make it possible for the user to interact with the system as a whole in its current primitive state.

Annex B
Basic System Diagrams

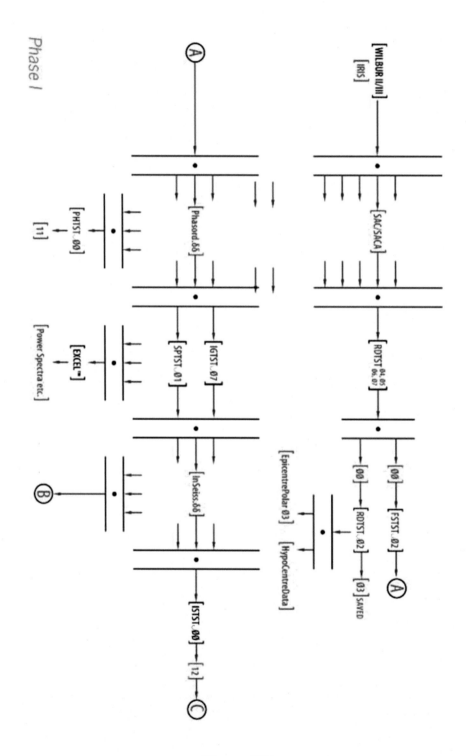

Figure B01

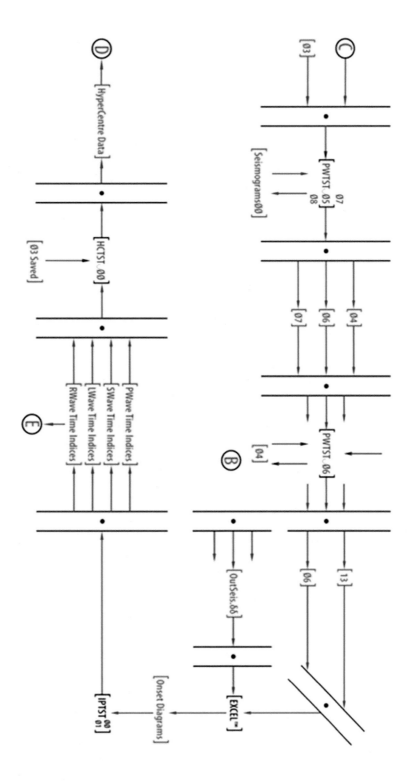

Figure B02

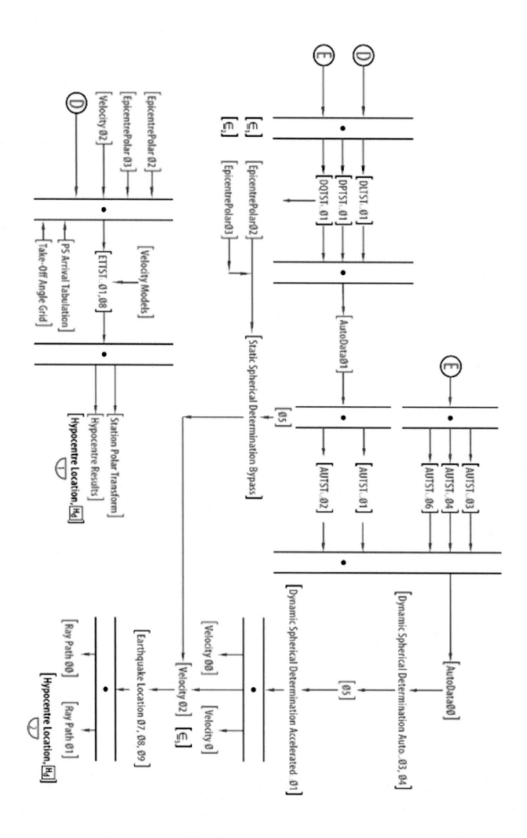

Figure B03

Annex C

The Programs UETST 00.cpp and UETST 01.cpp

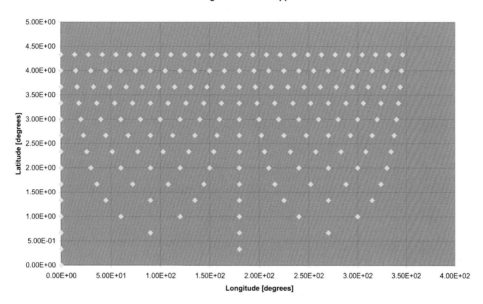

Figure C01

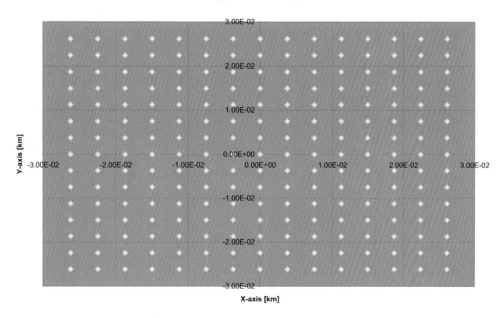

Figure C02

Lat/Long Coordinates of Rectangular Grid of Epicentres
Program UETST 01.cpp

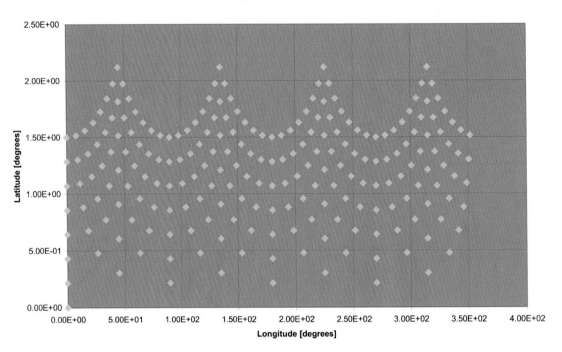

Figure C03

Scan of 7-ring Azimuthal Grid
Central Epicenter: Lat/Long = 46.637N/151.136E
Increment in Relative Co-Latitude 0.333333 degrees
Interpolation Method: Linear/Cubic
H[d], on Minimum RMS Residual, 120.145 km
Revised Epicenter: Lat/Long = 46.353N/151.136E

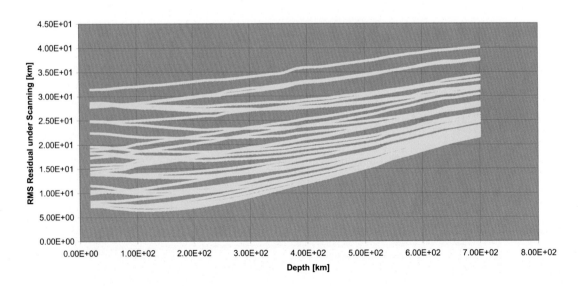

Figure C04

Annex D

Wave Structure by Frequency Band Splitting

(Some Results from the Routine IGTST 05.cpp)

Annex D shows how some of the elements of the set of information making up a waveform can be individually isolated and depicted as waveforms in their own right.

If we take the accelerogram containing all frequencies given in figure D00; perform an FFT; and integrate within the frequency domain, then we have a set (or list) of coefficients (due to the Fourier transform) equal to the number of sample points considered in the original input to the FFT.

In the frequency domain, we can integrate according to the scheme

$$\iint_{t_i} f(\tau) \cdot d\tau \leftrightarrow \left(\frac{1}{j\omega} \right)^2 F(\omega)$$

Here, the function $f(t_i)$ represents the entire accelerogram, sampled at the t_i, and $F(\omega)$ is its Fourier transform.

Ultimately, we can then slide or float the given bandpass window up the set of frequencies re-synthesising the integrated accelerogram from those coefficients that lie within this sliding window, as this window moves.

This sliding motion can be implemented by providing an arbitrary (or dynamic) step distance, for positioning the base of the bandpass window or interval as it moves down the set of coefficients. Some results, along the lines of this projection, are appended in this annex. This series represents integration, at a sample interval of 0.05 seconds, by the routine **IGTST 05**.

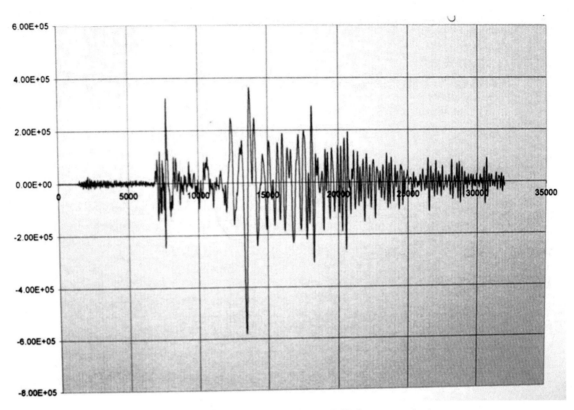

Figure D00 – Accelerogram (all frequencies)

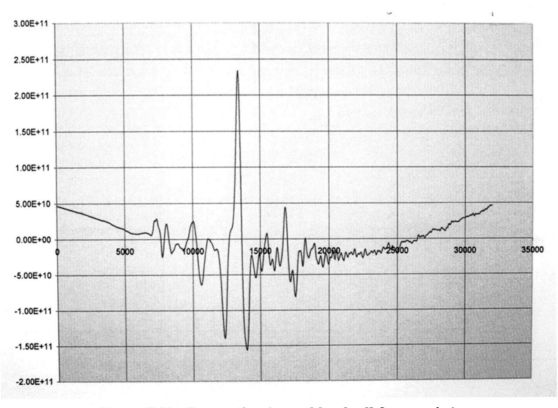

Figure D01 – Integration (second level; all frequencies)

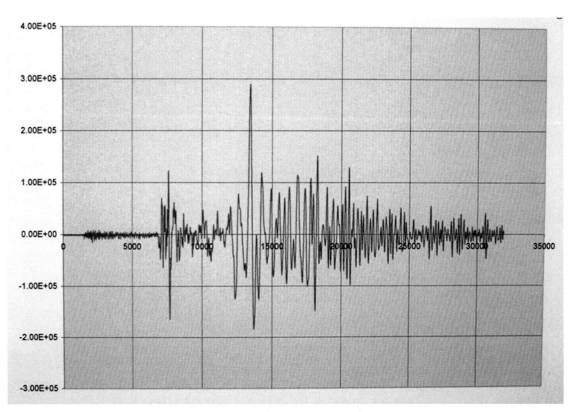

Figure D02 – Integration (second level; 5 to 10 Hz)

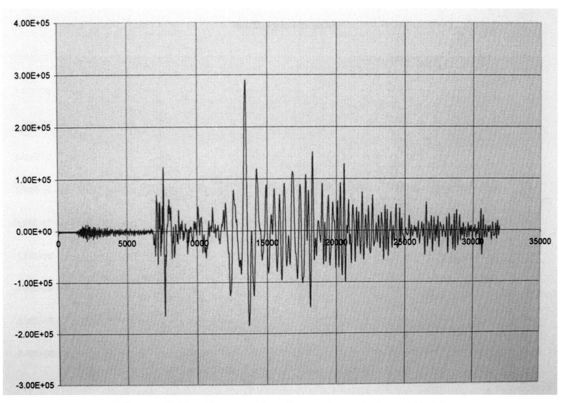

Figure D03 – Integration (second level; 2 to 10 Hz)

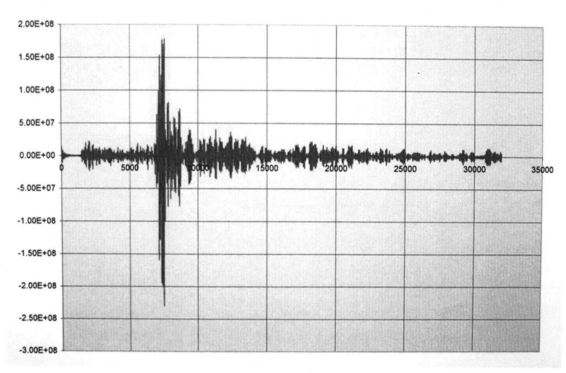

Figure D04 – Integration (2nd level; 0.1 to 1.0Hz)

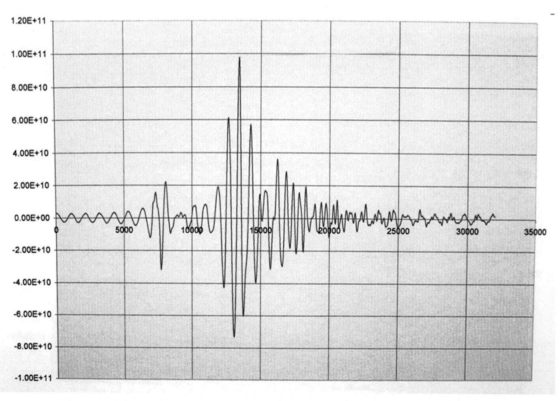

Figure D05 – Integration (second level; 0.01 to 0.1 Hz)

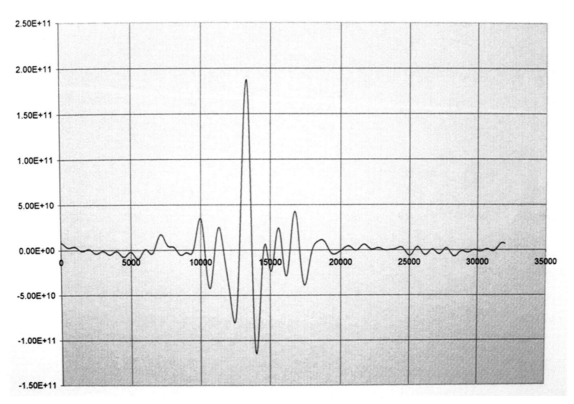

Figure D06 – Integration (second level; 0.001 to 0.01 Hz)

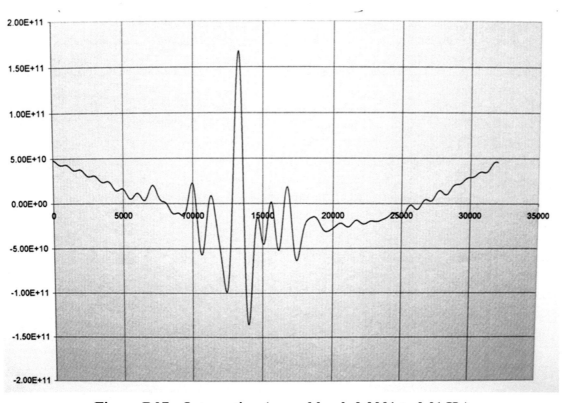

Figure D07 – Integration (second level; 0.0001 to 0.01 Hz)

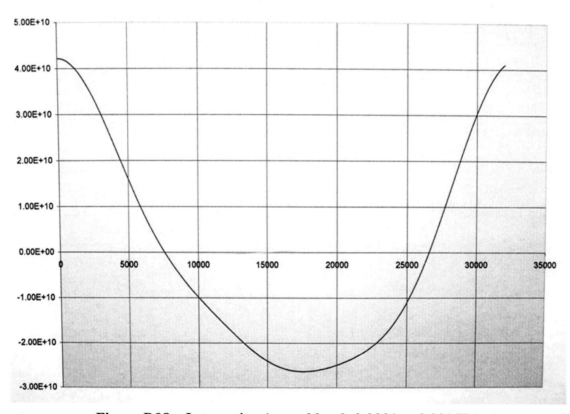

Figure D08 – Integration (second level; 0.0001 to 0.001 Hz)

Annex E

Levels of Integration: Verification of FFT against IDFT

Annex E shows how the integration techniques mentioned in the main text—DFT, IDFT, and FFT—yield identical results.

The IDFT (inverse discrete Fourier transform) method of integration, as implemented in IGTST 04, is compared with the use of the FFT (fast Fourier transform) routine, as implemented in IGTST 05, to perform the same set of levels of integration.

These levels are:

- Level 00: *Re-synthesising* the original function (for example, accelerogram)
- Level 01: *Velocity* integrated from Level 00
- Level 02: *Movement* integrated from Level 00
- Level 03: *Displacement* integrated from Level 00

The frequency band used for all integrations was "all frequencies".
This situation is demonstrated in the figures included in this annex.

00: IGTST 04 [Accelerogram]

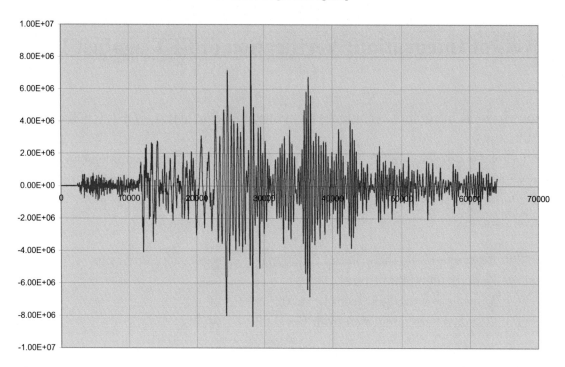

Figure E00

00: IGTST 05 [Accelerogram]

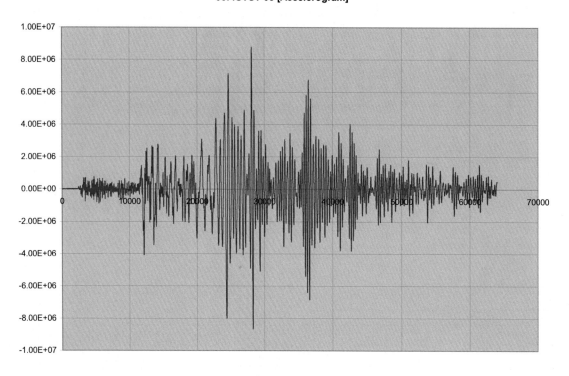

Figure E01

00: IGTST 04 [Velocity]

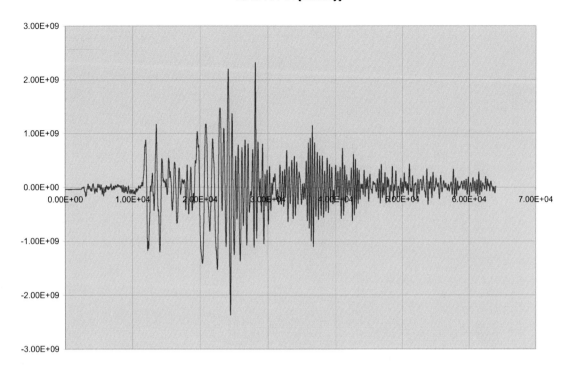

Figure E02

00: IGTST 05 [Velocity]

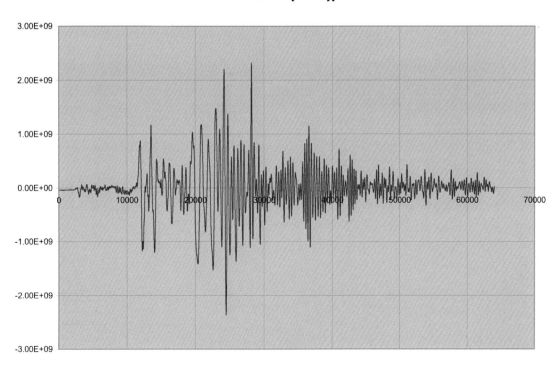

Figure E03

00: IGTST 04 [Motion]

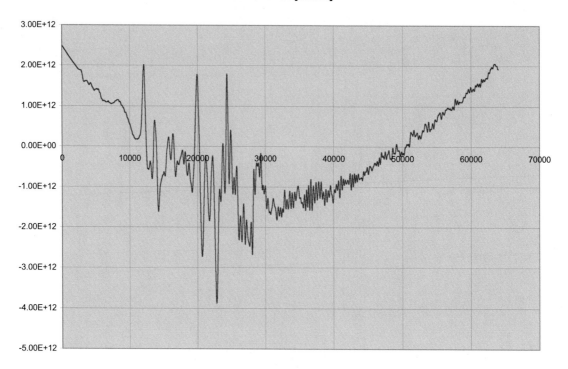

Figure E04

00: IGTST 05 [Motion]

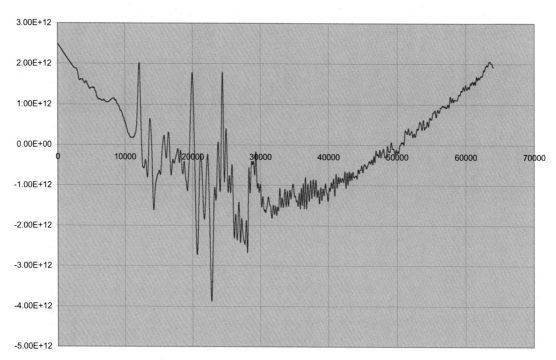

Figure E05

00: IGTST 04 [Displacement]

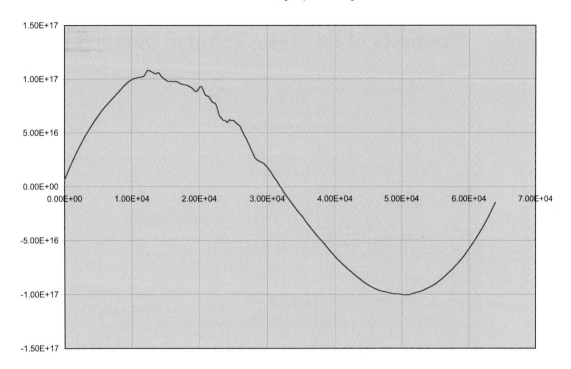

Figure E06

00: IGTST 05 [Displacement]

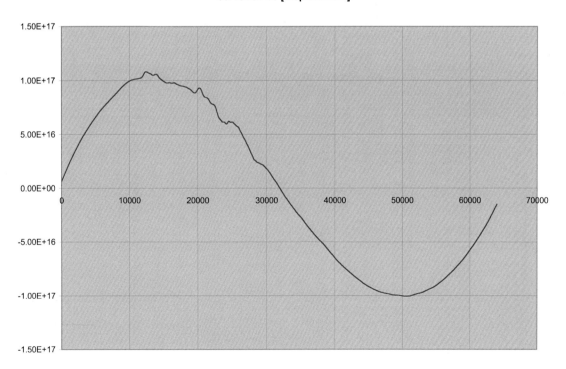

Figure E07

Annex F

Examples of the Basic Tabular Data

Hypocentre Location "Lagrange" [Delta Theta = Constant]
[ak135]
Travel Time Database [51 by 51]
P-wave: Uninterpolated

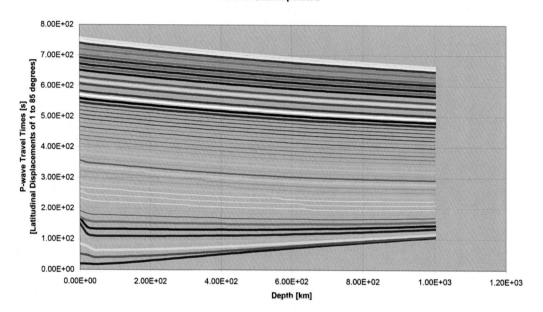

Figure F01

Hypocentre Location "Lagrange" [Delta Theta = Constant]
[ak135]
Take-off Angle Database [51 by 51]
P-wave: Uninterpolated

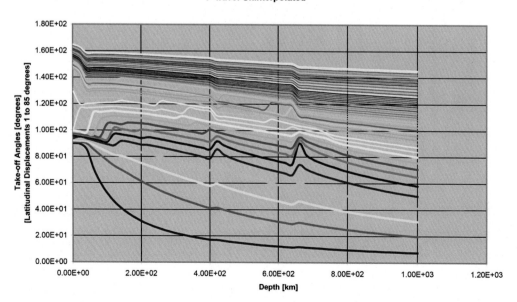

Figure F02

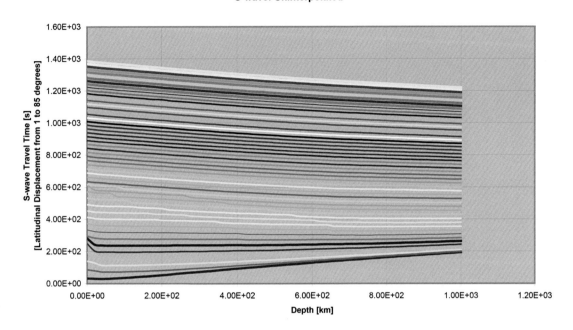

Hypocentre Location "Lagrange" [Delta Theta = Constant]
[ak135]
Travel Time Database [51 by 51]
S-wave: Uninterpolated

Figure F03

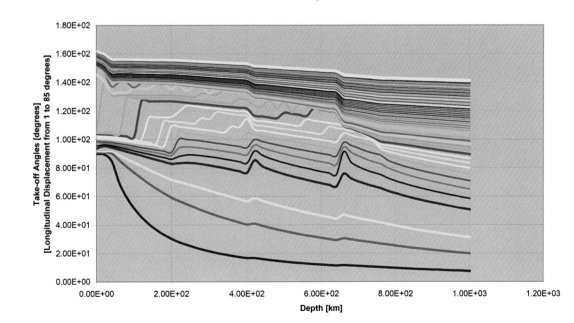

Hypocentre Location "Lagrange" [DeltaTheta = Constant]
[ak135]
Take-off Angle Database [51 by 51]
S-wave: Uninterpolated

Figure F04

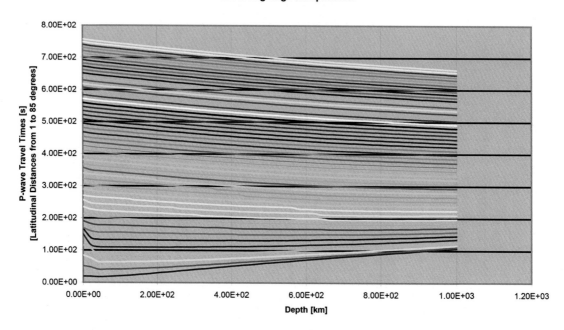

Figure F05

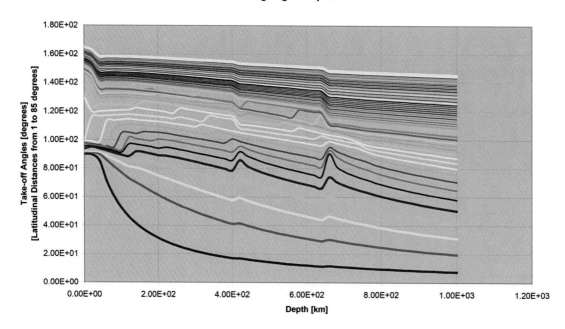

Figure F06

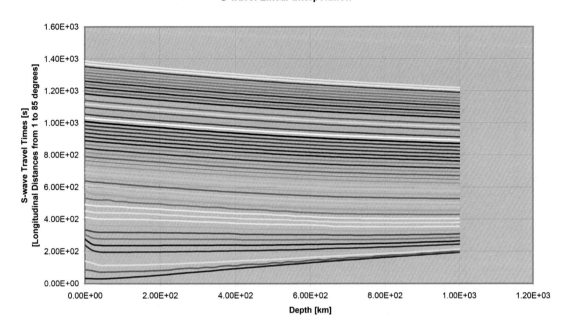

Figure F07

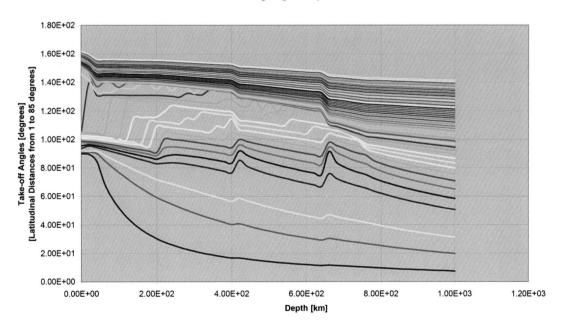

Figure F08

Annex G

Some Preliminary Results of Hypocentre Determination

Data Derived from Seismograms

Stn.	Seismogram Initial Point (s)	Basal Increment (s)	P-Wave Arrival (s)	Bassal Time Index
	(1)	(2)	(3)	(4)
1.	1.0788287e+004	1.4885000e+002	1.0937137e+004	2.9770000e+003
2.	1.0829205e+004	1.3975000e+002	1.0968955e+004	2.7950000e+003
3.	1.0831169e+004	1.2980000e+002	1.0960969e+004	2.5960000e+003
4.	1.0879669e+004	1.1820000e+002	1.0997869e+004	2.3640000e+003
5.	1.0934248e+004	1.2815000e+002	1.1062398e+004	2.5630000e+003
6.	1.1064669e+004	9.2450001e+001	1.1157119e+004	1.8490000e+003
7.	1.1074269e+004	7.5850001e+001	1.1150119e+004	1.5170000e+003
8.	1.1111635e+004	8.9950001e+001	1.1201585e+004	1.7990000e+003
9.	1.1116019e+004	8.1700001e+001	1.1197719e+004	1.6340000e+003
10.	1.1130569e+004	9.4100001e+001	1.1224669e+004	1.8820000e+003
11.	1.1146648e+004	8.1700001e+001	1.1228348e+004	1.6340000e+003
12.	1.1154219e+004	9.5800001e+001	1.1250019e+004	1.9160000e+003
13.	1.1163069e+004	8.4150001e+001	1.1247219e+004	1.6830000e+003
15.	1.1171069e+004	8.1700001e+001	1.1252769e+004	1.6340000e+003
17.	1.1200619e+004	8.0850001e+001	1.1281469e+004	1.6170000e+003
18.	1.1225869e+004	9.4100001e+001	1.1319969e+004	1.8820000e+003
19.	1.1243419e+004	7.5850001e+001	1.1319269e+004	1.5170000e+003
20.	1.1245469e+004	8.1700001e+001	1.1327169e+004	1.6340000e+003
21.	1.1396019e+004	8.0000001e+001	1.1476019e+004	1.6000000e+003
22.	1.1390619e+004	6.7600001e+001	1.1458219e+004	1.3520000e+003
23.	1.1386969e+004	6.6750001e+001	1.1453719e+004	1.3350000e+003
24.	1.1379569e+004	7.5850001e+001	1.1455419e+004	1.5170000e+003
25.	1.1378019e+004	7.8350001e+001	1.1456369e+004	1.5670000e+003
26.	1.1346869e+004	7.0050001e+001	1.1416919e+004	1.4010000e+003
27.	1.1314219e+004	7.3400001e+001	1.1387619e+004	1.4680000e+003
28.	1.1280560e+004	6.8400001e+001	1.1348960e+004	1.3680000e+003

The P-wave arrival at each station, taken as active, is calculated as:

$$(3) = (1) + (2)$$

$$(2) = \Delta t \times (4),$$

Δt being the sample interval, in this case set at 0.05s.

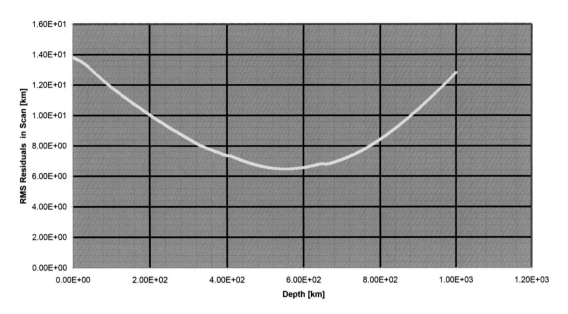

Hypocentre Location [Radial] ak135
Table Formation: Linear
Interpolation Method: Linear/Linear
H[d] = 556.5 km
[Consensus H[d] = 556.5 km]

Figure G01

```
        Station  1: Take Off Angle 8.7640908e+001 degrees.
        Station  2: Take Off Angle 9.6883222e+001 degrees.
        Station  3: Take Off Angle 9.4925307e+001 degrees.
        Station  4: Take Off Angle 1.0406762e+002 degrees.
        Station  5: Take Off Angle 1.2097129e+002 degrees.
        Station  6: Take Off Angle 1.2708507e+002 degrees.
        Station  7: Take Off Angle 1.2703163e+002 degrees.
        Station  8: Take Off Angle 1.3019376e+002 degrees.
        Station  9: Take Off Angle 1.2983421e+002 degrees.
        Station 10: Take Off Angle 1.3189769e+002 degrees.
        Station 11: Take Off Angle 1.3205546e+002 degrees.
        Station 12: Take Off Angle 1.3378032e+002 degrees.
        Station 13: Take Off Angle 1.3355505e+002 degrees.
        Station 15: Take Off Angle 1.3400222e+002 degrees.
        Station 17: Take Off Angle 1.3612695e+002 degrees.
        Station 18: Take Off Angle 1.3777107e+002 degrees.
        Station 19: Take Off Angle 1.3930410e+002 degrees.
        Station 20: Take Off Angle 1.3962629e+002 degrees.
        Station 21: Take Off Angle 1.5074713e+002 degrees.
        Station 22: Take Off Angle 1.5008385e+002 degrees.
        Station 23: Take Off Angle 1.5060541e+002 degrees.
        Station 24: Take Off Angle 1.5013796e+002 degrees.
        Station 25: Take Off Angle 1.4922030e+002 degrees.
        Station 26: Take Off Angle 1.4704728e+002 degrees.
        Station 27: Take Off Angle 1.4398287e+002 degrees.
        Station 28: Take Off Angle 1.4284324e+002 degrees.
```

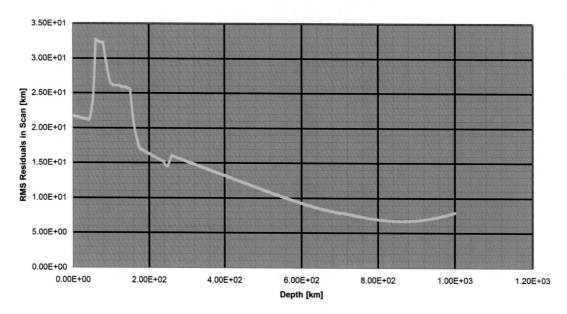

Figure G02

```
Station  1: Take Off Angle 7.4037975e+001 degrees.
Station  2: Take Off Angle 8.3066038e+001 degrees.
Station  3: Take Off Angle 8.1202156e+001 degrees.
Station  4: Take Off Angle 8.9546876e+001 degrees.
Station  5: Take Off Angle 9.8960358e+001 degrees.
Station  6: Take Off Angle 1.1099772e+002 degrees.
Station  7: Take Off Angle 1.1090573e+002 degrees.
Station  8: Take Off Angle 1.1605273e+002 degrees.
Station  9: Take Off Angle 1.1556562e+002 degrees.
Station 10: Take Off Angle 1.1882618e+002 degrees.
Station 11: Take Off Angle 1.1905702e+002 degrees.
Station 12: Take Off Angle 1.2177295e+002 degrees.
Station 13: Take Off Angle 1.2144766e+002 degrees.
Station 15: Take Off Angle 1.2208302e+002 degrees.
Station 17: Take Off Angle 1.2532824e+002 degrees.
Station 18: Take Off Angle 1.2775541e+002 degrees.
Station 19: Take Off Angle 1.3018952e+002 degrees.
Station 20: Take Off Angle 1.3064841e+002 degrees.
Station 21: Take Off Angle 1.4611786e+002 degrees.
Station 22: Take Off Angle 1.4533563e+002 degrees.
Station 23: Take Off Angle 1.4593679e+002 degrees.
Station 24: Take Off Angle 1.4539625e+002 degrees.
Station 25: Take Off Angle 1.4422500e+002 degrees.
Station 26: Take Off Angle 1.4128099e+002 degrees.
Station 27: Take Off Angle 1.3711340e+002 degrees.
Station 28: Take Off Angle 1.3546074e+002 degrees.
```

"Radial" Scanning Method(3): Hypocentre Focal Depth: 5.515000e+002 km

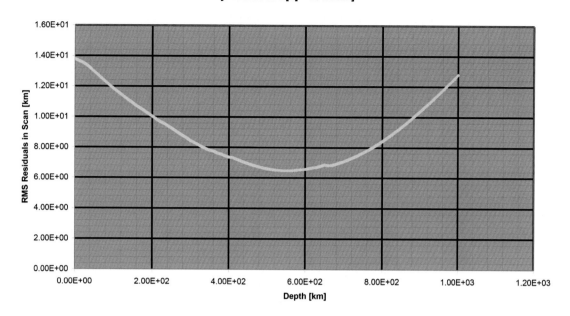

Figure G03

```
Station  1: Take Off Angle 8.8007821e+001 degrees.
Station  2: Take Off Angle 9.6932953e+001 degrees.
Station  3: Take Off Angle 9.5398472e+001 degrees.
Station  4: Take Off Angle 1.0383435e+002 degrees.
Station  5: Take Off Angle 1.2116253e+002 degrees.
Station  6: Take Off Angle 1.2711216e+002 degrees.
Station  7: Take Off Angle 1.2703985e+002 degrees.
Station  8: Take Off Angle 1.3034969e+002 degrees.
Station  9: Take Off Angle 1.3008341e+002 degrees.
Station 10: Take Off Angle 1.3202974e+002 degrees.
Station 11: Take Off Angle 1.3214061e+002 degrees.
Station 12: Take Off Angle 1.3397405e+002 degrees.
Station 13: Take Off Angle 1.3383686e+002 degrees.
Station 15: Take Off Angle 1.3411628e+002 degrees.
Station 17: Take Off Angle 1.3634886e+002 degrees.
Station 18: Take Off Angle 1.3790103e+002 degrees.
Station 19: Take Off Angle 1.3933541e+002 degrees.
Station 20: Take Off Angle 1.3974939e+002 degrees.
Station 21: Take Off Angle 1.5099127e+002 degrees.
Station 22: Take Off Angle 1.4989314e+002 degrees.
Station 23: Take Off Angle 1.5092660e+002 degrees.
Station 24: Take Off Angle 1.5002167e+002 degrees.
Station 25: Take Off Angle 1.4939982e+002 degrees.
Station 26: Take Off Angle 1.4716174e+002 degrees.
Station 27: Take Off Angle 1.4406544e+002 degrees.
Station 28: Take Off Angle 1.4300802e+002 degrees.
```

"Radial" Scanning Method(4): Hypocentre Focal Depth: 8.633760e+002 km

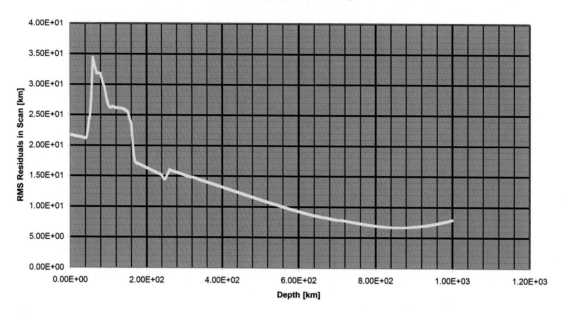

Figure G04

```
    Station  1: Take Off Angle 7.4109462e+001 degrees.
    Station  2: Take Off Angle 8.3093903e+001 degrees.
    Station  3: Take Off Angle 8.1239548e+001 degrees.
    Station  4: Take Off Angle 8.9483283e+001 degrees.
    Station  5: Take Off Angle 9.8788738e+001 degrees.
    Station  6: Take Off Angle 1.1106048e+002 degrees.
    Station  7: Take Off Angle 1.1097260e+002 degrees.
    Station  8: Take Off Angle 1.1604630e+002 degrees.
    Station  9: Take Off Angle 1.1560166e+002 degrees.
    Station 10: Take Off Angle 1.1882047e+002 degrees.
    Station 11: Take Off Angle 1.1901336e+002 degrees.
    Station 12: Take Off Angle 1.2164539e+002 degrees.
    Station 13: Take Off Angle 1.2139443e+002 degrees.
    Station 15: Take Off Angle 1.2197729e+002 degrees.
    Station 17: Take Off Angle 1.2527167e+002 degrees.
    Station 18: Take Off Angle 1.2795767e+002 degrees.
    Station 19: Take Off Angle 1.3020994e+002 degrees.
    Station 20: Take Off Angle 1.3062682e+002 degrees.
    Station 21: Take Off Angle 1.4602505e+002 degrees.
    Station 22: Take Off Angle 1.4530959e+002 degrees.
    Station 23: Take Off Angle 1.4589921e+002 degrees.
    Station 24: Take Off Angle 1.4537414e+002 degrees.
    Station 25: Take Off Angle 1.4421407e+002 degrees.
    Station 26: Take Off Angle 1.4138152e+002 degrees.
    Station 27: Take Off Angle 1.3718569e+002 degrees.
    Station 28: Take Off Angle 1.3539890e+002 degrees.
```

216

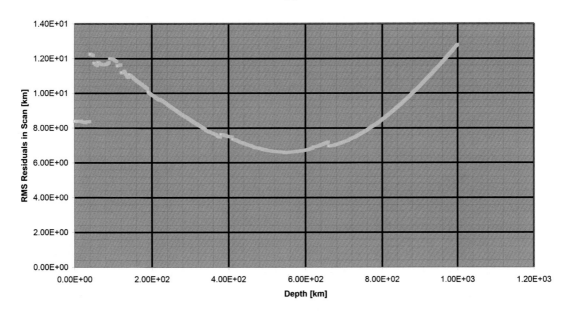

Figure G05

```
Station 1: Take Off Angle 8.5345416e+001 degrees.
Station 2: Take Off Angle 9.4869161e+001 degrees.
Station 3: Take Off Angle 9.2769079e+001 degrees.
Station 4: Take Off Angle 1.0259748e+002 degrees.
Station 5: Take Off Angle 1.2131622e+002 degrees.
Station 6: Take Off Angle 1.2124458e+002 degrees.
Station 7: Take Off Angle 1.2524427e+002 degrees.
Station 8: Take Off Angle 1.2483730e+002 degrees.
Station 9: Take Off Angle 1.2734720e+002 degrees.
Station 10: Take Off Angle 1.2754744e+002 degrees.
Station 11: Take Off Angle 1.2957608e+002 degrees.
Station 12: Take Off Angle 1.2931247e+002 degrees.
Station 13: Take Off Angle 1.2982736e+002 degrees.
Station 15: Take Off Angle 1.3235552e+002 degrees.
Station 17: Take Off Angle 1.3421696e+002 degrees.
Station 18: Take Off Angle 1.3595259e+002 degrees.
Station 19: Take Off Angle 1.3632612e+002 degrees.
Station 20: Take Off Angle 1.4857221e+002 degrees.
Station 21: Take Off Angle 1.4792192e+002 degrees.
Station 22: Take Off Angle 1.4842697e+002 degrees.
Station 23: Take Off Angle 1.4797353e+002 degrees.
Station 24: Take Off Angle 1.4695783e+002 degrees.
Station 25: Take Off Angle 1.4456901e+002 degrees.
Station 26: Take Off Angle 1.4125017e+002 degrees.
Station 27: Take Off Angle 1.3997257e+002 degrees.
```

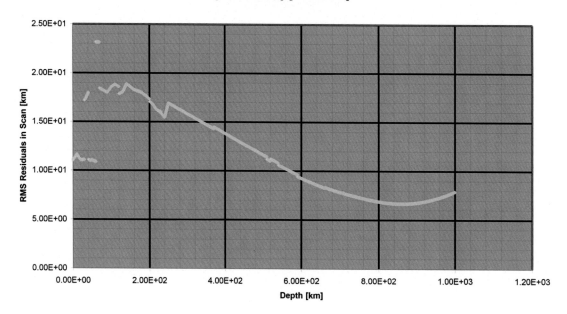

Figure G06

```
Station 1: Take Off Angle 7.0317219e+001 degrees.
Station 2: Take Off Angle 7.9486837e+001 degrees.
Station 3: Take Off Angle 7.7526735e+001 degrees.
Station 4: Take Off Angle 8.6316844e+001 degrees.
Station 5: Take Off Angle 9.6860021e+001 degrees.
Station 6: Take Off Angle 1.0993225e+002 degrees.
Station 7: Take Off Angle 1.0983868e+002 degrees.
Station 8: Take Off Angle 1.1520373e+002 degrees.
Station 9: Take Off Angle 1.1464987e+002 degrees.
Station 10: Take Off Angle 1.1796794e+002 degrees.
Station 11: Take Off Angle 1.1824372e+002 degrees.
Station 12: Take Off Angle 1.2096146e+002 degrees.
Station 13: Take Off Angle 1.2060095e+002 degrees.
Station 15: Take Off Angle 1.2130113e+002 degrees.
Station 17: Take Off Angle 1.2467434e+002 degrees.
Station 18: Take Off Angle 1.2716976e+002 degrees.
Station 19: Take Off Angle 1.2957637e+002 degrees.
Station 20: Take Off Angle 1.3004307e+002 degrees.
Station 21: Take Off Angle 1.4585971e+002 degrees.
Station 22: Take Off Angle 1.4509297e+002 degrees.
Station 23: Take Off Angle 1.4569353e+002 degrees.
Station 24: Take Off Angle 1.4515498e+002 degrees.
Station 25: Take Off Angle 1.4391994e+002 degrees.
Station 26: Take Off Angle 1.4095832e+002 degrees.
Station 27: Take Off Angle 1.3668262e+002 degrees.
Station 28: Take Off Angle 1.3496109e+002 degrees.
```

218

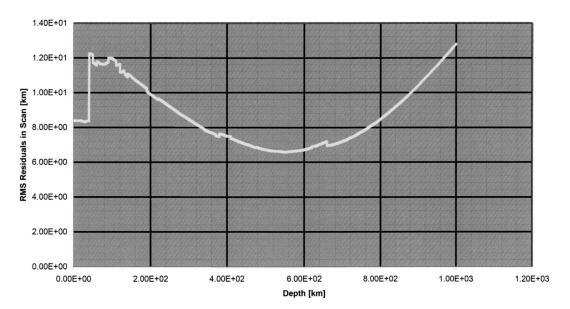

Figure G07

```
Station  1: Take Off Angle 8.5345530e+001 degrees.
Station  2: Take Off Angle 9.4869337e+001 degrees.
Station  3: Take Off Angle 9.2769087e+001 degrees.
Station  4: Take Off Angle 1.0259741e+002 degrees.
Station  5: Take Off Angle 1.2131602e+002 degrees.
Station  6: Take Off Angle 1.2124089e+002 degrees.
Station  7: Take Off Angle 1.2524429e+002 degrees.
Station  8: Take Off Angle 1.2483731e+002 degrees.
Station  9: Take Off Angle 1.2734720e+002 degrees.
Station 10: Take Off Angle 1.2754744e+002 degrees.
Station 11: Take Off Angle 1.2957608e+002 degrees.
Station 12: Take Off Angle 1.2931248e+002 degrees.
Station 13: Take Off Angle 1.2982736e+002 degrees.
Station 15: Take Off Angle 1.3235554e+002 degrees.
Station 17: Take Off Angle 1.3422043e+002 degrees.
Station 18: Take Off Angle 1.3595259e+002 degrees.
Station 19: Take Off Angle 1.3632613e+002 degrees.
Station 20: Take Off Angle 1.4857223e+002 degrees.
Station 21: Take Off Angle 1.4792193e+002 degrees.
Station 22: Take Off Angle 1.4842699e+002 degrees.
Station 23: Take Off Angle 1.4797354e+002 degrees.
Station 24: Take Off Angle 1.4695781e+002 degrees.
Station 25: Take Off Angle 1.4456900e+002 degrees.
Station 26: Take Off Angle 1.4125017e+002 degrees.
Station 27: Take Off Angle 1.3996910e+002 degrees.
```

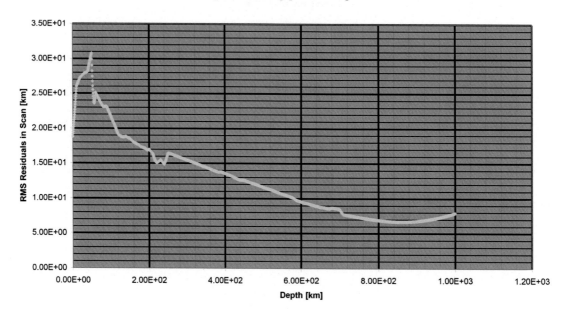

Figure G08

```
Station  1: Take Off Angle 7.0317143e+001 degrees.
Station  2: Take Off Angle 7.9486797e+001 degrees.
Station  3: Take Off Angle 7.7526675e+001 degrees.
Station  4: Take Off Angle 8.6316809e+001 degrees.
Station  5: Take Off Angle 9.6860079e+001 degrees.
Station  6: Take Off Angle 1.0993227e+002 degrees.
Station  7: Take Off Angle 1.0983870e+002 degrees.
Station  8: Take Off Angle 1.1520373e+002 degrees.
Station  9: Take Off Angle 1.1464989e+002 degrees.
Station 10: Take Off Angle 1.1796794e+002 degrees.
Station 11: Take Off Angle 1.1824370e+002 degrees.
Station 12: Take Off Angle 1.2096146e+002 degrees.
Station 13: Take Off Angle 1.2060095e+002 degrees.
Station 15: Take Off Angle 1.2130113e+002 degrees.
Station 17: Take Off Angle 1.2467433e+002 degrees.
Station 18: Take Off Angle 1.2716975e+002 degrees.
Station 19: Take Off Angle 1.2957635e+002 degrees.
Station 20: Take Off Angle 1.3004305e+002 degrees.
Station 21: Take Off Angle 1.4585972e+002 degrees.
Station 22: Take Off Angle 1.4509293e+002 degrees.
Station 23: Take Off Angle 1.4569354e+002 degrees.
Station 24: Take Off Angle 1.4515495e+002 degrees.
Station 25: Take Off Angle 1.4391993e+002 degrees.
Station 26: Take Off Angle 1.4095829e+002 degrees.
Station 27: Take Off Angle 1.3668263e+002 degrees.
Station 28: Take Off Angle 1.3496110e+002 degrees.
```

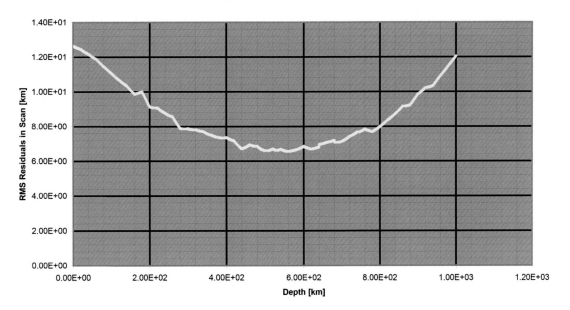

Figure G09

```
    Station  1: Take Off Angle 8.7707239e+001 degrees.
    Station  2: Take Off Angle 9.7236470e+001 degrees.
    Station  3: Take Off Angle 9.5325414e+001 degrees.
    Station  4: Take Off Angle 1.0426348e+002 degrees.
    Station  5: Take Off Angle 1.2096160e+002 degrees.
    Station  6: Take Off Angle 1.2713319e+002 degrees.
    Station  7: Take Off Angle 1.2707838e+002 degrees.
    Station  8: Take Off Angle 1.3031429e+002 degrees.
    Station  9: Take Off Angle 1.2991286e+002 degrees.
    Station 10: Take Off Angle 1.3204196e+002 degrees.
    Station 11: Take Off Angle 1.3219039e+002 degrees.
    Station 12: Take Off Angle 1.3388775e+002 degrees.
    Station 13: Take Off Angle 1.3368046e+002 degrees.
    Station 15: Take Off Angle 1.3410401e+002 degrees.
    Station 17: Take Off Angle 1.3623477e+002 degrees.
    Station 18: Take Off Angle 1.3789209e+002 degrees.
    Station 19: Take Off Angle 1.3940621e+002 degrees.
    Station 20: Take Off Angle 1.3967696e+002 degrees.
    Station 21: Take Off Angle 1.5091974e+002 degrees.
    Station 22: Take Off Angle 1.5032209e+002 degrees.
    Station 23: Take Off Angle 1.5077179e+002 degrees.
    Station 24: Take Off Angle 1.5036620e+002 degrees.
    Station 25: Take Off Angle 1.4935193e+002 degrees.
    Station 26: Take Off Angle 1.4713608e+002 degrees.
    Station 27: Take Off Angle 1.4416873e+002 degrees.
    Station 28: Take Off Angle 1.4300838e+002 degrees.
```

221

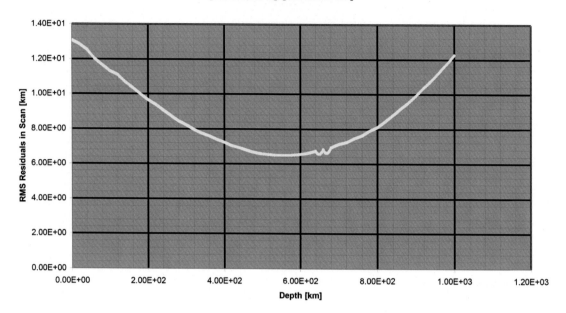

Figure G10

```
    Station 1: Take Off Angle 8.8324561e+001 degrees.
    Station 2: Take Off Angle 9.7447189e+001 degrees.
    Station 3: Take Off Angle 9.5500535e+001 degrees.
    Station 4: Take Off Angle 1.0448830e+002 degrees.
    Station 5: Take Off Angle 1.2149588e+002 degrees.
    Station 6: Take Off Angle 1.2745351e+002 degrees.
    Station 7: Take Off Angle 1.2740167e+002 degrees.
    Station 8: Take Off Angle 1.3059281e+002 degrees.
    Station 9: Take Off Angle 1.3025408e+002 degrees.
   Station 10: Take Off Angle 1.3228274e+002 degrees.
   Station 11: Take Off Angle 1.3245212e+002 degrees.
   Station 12: Take Off Angle 1.3414687e+002 degrees.
   Station 13: Take Off Angle 1.3391774e+002 degrees.
   Station 15: Take Off Angle 1.3436232e+002 degrees.
   Station 17: Take Off Angle 1.3649673e+002 degrees.
   Station 18: Take Off Angle 1.3808619e+002 degrees.
   Station 19: Take Off Angle 1.3964137e+002 degrees.
   Station 20: Take Off Angle 1.3994944e+002 degrees.
   Station 21: Take Off Angle 1.5103157e+002 degrees.
   Station 22: Take Off Angle 1.5041279e+002 degrees.
   Station 23: Take Off Angle 1.5089632e+002 degrees.
   Station 24: Take Off Angle 1.5046257e+002 degrees.
   Station 25: Take Off Angle 1.4953278e+002 degrees.
   Station 26: Take Off Angle 1.4731953e+002 degrees.
   Station 27: Take Off Angle 1.4433965e+002 degrees.
   Station 28: Take Off Angle 1.4316282e+002 degrees.
```

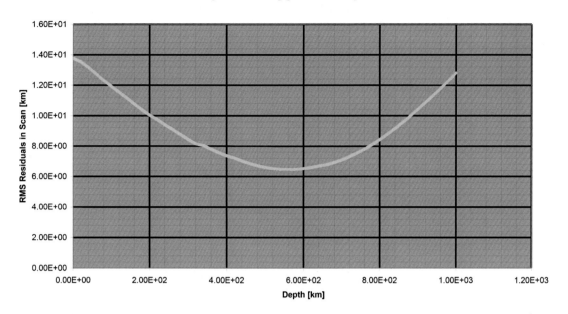

Hypocentre Location [Lagrange] ak135
Table preparation: Linear
Interpolation Method: Linear/Linear
H[d] = 571.5 km
[Consensus H[d] = 572.250 km]

Figure G11

```
Station  1: Take Off Angle 8.6519650e+001 degrees.
Station  2: Take Off Angle 9.9039326e+001 degrees.
Station  3: Take Off Angle 9.5736070e+001 degrees.
Station  4: Take Off Angle 1.1062719e+002 degrees.
Station  5: Take Off Angle 1.2026741e+002 degrees.
Station  6: Take Off Angle 1.2647591e+002 degrees.
Station  7: Take Off Angle 1.2642146e+002 degrees.
Station  8: Take Off Angle 1.2966634e+002 degrees.
Station  9: Take Off Angle 1.2932280e+002 degrees.
Station 10: Take Off Angle 1.3139143e+002 degrees.
Station 11: Take Off Angle 1.3156519e+002 degrees.
Station 12: Take Off Angle 1.3329994e+002 degrees.
Station 13: Take Off Angle 1.3306748e+002 degrees.
Station 15: Take Off Angle 1.3351816e+002 degrees.
Station 17: Take Off Angle 1.3569100e+002 degrees.
Station 18: Take Off Angle 1.3730505e+002 degrees.
Station 19: Take Off Angle 1.3888150e+002 degrees.
Station 20: Take Off Angle 1.3918930e+002 degrees.
Station 21: Take Off Angle 1.5041623e+002 degrees.
Station 22: Take Off Angle 1.4980080e+002 degrees.
Station 23: Take Off Angle 1.5028057e+002 degrees.
Station 24: Take Off Angle 1.4985005e+002 degrees.
Station 25: Take Off Angle 1.4888731e+002 degrees.
Station 26: Take Off Angle 1.4667790e+002 degrees.
Station 27: Take Off Angle 1.4364916e+002 degrees.
Station 28: Take Off Angle 1.4246331e+002 degrees.
```

In the above figures, in Radial Scanning Method (1), data tables generated by the radial ray tracer using the velocity model ak135 were used in the scan, and the tables were finalised with a linear interpolation.

In Radial Scanning Method (2), data tables generated by the radial ray tracer using the velocity model PREM were used in the scan, and the tables were finalised with a linear interpolation.

In Radial Scanning Method (3), data tables generated by the radial ray tracer using the velocity model ak135 were used in the scan, and the tables were finalised with a Lagrange interpolation.

In Radial Scanning Method (4), data tables generated by the radial ray tracer using the velocity model "PREM" were used in the scan, and the tables were finalised with a Lagrange interpolation.

In Eikonal Scanning Method (5), data tables generated by the eikonal ray tracer using the velocity model ak135 were used in the scan, and no finalisation of the tables took place in this case.

In Eikonal Scanning Method (6), data tables generated by the eikonal ray tracer using the velocity model PREM were used in the scan, and no finalisation of the tables took place in this case.

In Eikonal Scanning Method (7), data tables generated by the eikonal ray tracer using the velocity model ak135 were used in the scan, and the tables were finalised with a Lagrange interpolation.

In Eikonal Scanning Method (8), data tables generated by the eikonal ray tracer using the velocity model PREM were used in the scan, and the tables were finalised with a Lagrange interpolation.

In Radial Scanning Method (9), data tables generated by the radial ray tracer using the velocity model ak135 were used in the scan, and no finalisation of the tables took place in this case.

In Radial Scanning Method (10), data tables generated by the radial ray tracer using the velocity model ak135 were used in the scan, and the tables were finalised with a linear interpolation.

In Lagrange Scanning Method (11), data tables generated by the Lagrangian ray tracer using the velocity model ak135 were used in the scan, and the tables were finalised with a linear interpolation.

For all these methods, the depth scanned was from 1,000 kilometres with a granularity of 2,000 points. Total time for each individual scan was of the order of less than 0.25 seconds (Dell XPS600 at 3.2 gigaherz).

Annex H
Results of Second Trial of Tabular Scanning

Hypocentre Scan [Lagrange (4)] ak135 (P-Wave)
Table Normalisation Lagrange
Interpolation Method: Linear/Linear
H[d] = 113.670 km
Epicentre = [46.687N, 151.136E]

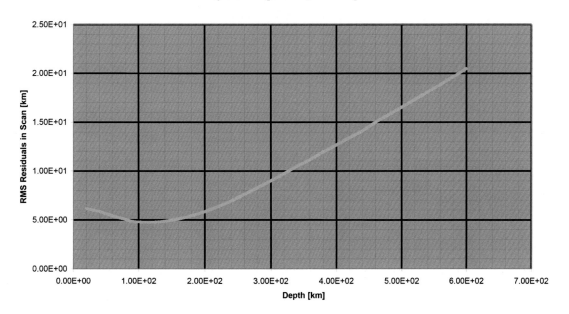

Figure H01

```
 Station  4: Take Off Angle 1.3739442e+002 degrees.
 Station  5: Take Off Angle 1.4377826e+002 degrees.
 Station  7: Take Off Angle 8.8958974e+001 degrees.
 Station  8: Take Off Angle 1.4079010e+002 degrees.
 Station 12: Take Off Angle 1.2917672e+002 degrees.
 Station 13: Take Off Angle 1.4140162e+002 degrees.
 Station 14: Take Off Angle 1.4030467e+002 degrees.
 Station 15: Take Off Angle 1.0418694e+002 degrees.
 Station 16: Take Off Angle 1.0746994e+002 degrees.
 Station 17: Take Off Angle 9.0400102e+001 degrees.
 Station 18: Take Off Angle 1.3637770e+002 degrees.
 Station 19: Take Off Angle 1.3936303e+002 degrees.
 Station 20: Take Off Angle 8.5312509e+001 degrees.
```

RMS Values (across the four interpolation methods)

```
4.7640837e+000   5.3206970e+000   5.0196899e+000   5.3278532e+000
```

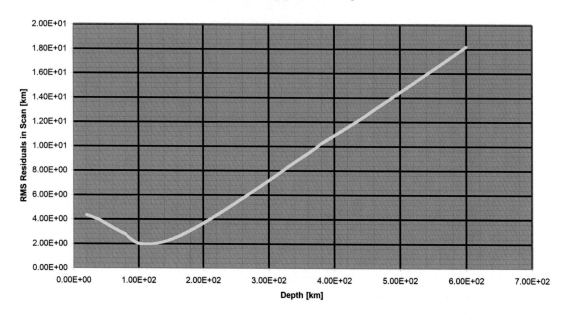

Hypocentre Scan [Lagrange (4)] ak135 (P-wave)
Table Normalisation Lagrange
Interpolation Method: Linear/Linear
H[d] = 112.220 km
[Consensus H[d] = 111.496 km]

Figure H02

```
Station 04: Take Off Angle 1.4376101e+002 degrees.
Station 05: Take Off Angle 1.4039191e+002 degrees.
Station 07: Take Off Angle 1.4095585e+002 degrees.
Station 08: Take Off Angle 1.4253667e+002 degrees.
Station 11: Take Off Angle 1.2998953e+002 degrees.
Station 12: Take Off Angle 1.4158699e+002 degrees.
Station 13: Take Off Angle 1.4016998e+002 degrees.
Station 14: Take Off Angle 1.0518575e+002 degrees.
Station 15: Take Off Angle 1.0560776e+002 degrees.
Station 16: Take Off Angle 9.1444576e+001 degrees.
Station 17: Take Off Angle 1.3701754e+002 degrees.
Station 18: Take Off Angle 1.3926449e+002 degrees.
Station 19: Take Off Angle 8.5083476e+001 degrees.
```

Final Station Residuals: (across the 4 Interpolation Methods).
```
Stn 04: 1.1166081e+000  1.0813991e+000  1.0799441e+000  1.0804342e+000
Stn 05: 1.5753146e+000  1.6078033e+000  1.6042875e+000  1.6037974e+000
Stn 07: 7.6804879e-001  7.3508588e-001  7.3248402e-001  7.3297407e-001
Stn 08: 3.4465880e+000  3.4808357e+000  3.4831561e+000  3.4826661e+000
Stn 11: 9.9965705e-001  1.0231965e+000  1.0545276e+000  1.0540376e+000
Stn 12: 2.6474382e-001  2.3124062e-001  2.1898664e-001  2.1947669e-001
Stn 13: 2.4433004e+000  2.4755625e+000  2.4885676e+000  2.4880776e+000
Stn 14: 3.2550137e+000  3.2362356e+000  3.1998725e+000  3.2003626e+000
Stn 15: 2.8831515e-001  2.7650796e-001  2.2955685e-001  2.3004690e-001
Stn 16: 7.6430232e-001  7.4975767e-001  6.9628632e-001  6.9579627e-001
Stn 17: 7.5990561e-002  3.1379229e-001  3.5858453e-001  3.5907459e-001
Stn 18: 3.9550819e+000  3.9234814e+000  3.9021963e+000  3.9026863e+000
Stn 19: 4.9463959e-001  4.6058722e-001  3.9479981e-001  4.0068045e-001
```

Hypocentre Scan [Lagrange (4)] ak135 (P-wave)
Table Normalisation Lagrange
Interpolation Method: Linear/Linear
H[d] = 111.640 km
[Consensus H[d] = 110.916 km]

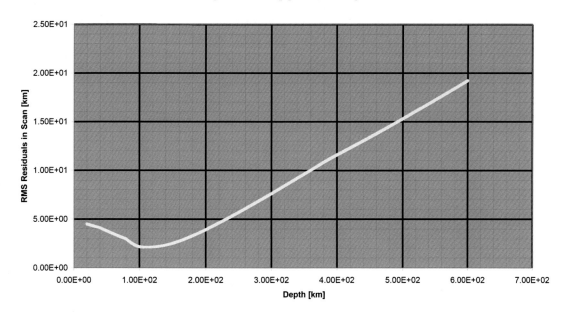

Figure H03

```
Station 04: Take Off Angle 1.4376385e+002 degrees.
Station 05: Take Off Angle 1.4039543e+002 degrees.
Station 07: Take Off Angle 1.4095911e+002 degrees.
Station 08: Take Off Angle 1.4253963e+002 degrees.
Station 12: Take Off Angle 1.4159015e+002 degrees.
Station 13: Take Off Angle 1.4017349e+002 degrees.
Station 14: Take Off Angle 1.0502058e+002 degrees.
Station 16: Take Off Angle 9.1378770e+001 degrees.
Station 17: Take Off Angle 1.3702105e+002 degrees.
Station 18: Take Off Angle 1.3927633e+002 degrees.
Station 19: Take Off Angle 8.5138013e+001 degrees.
```

Final Station Residuals (across the 4 Interpolation Methods):
```
Stn 04: 1.0370301e+000 9.9780745e-001 9.8950000e-001 9.9009192e-001
Stn 05: 1.6522566e+000 1.6887583e+000 1.6920951e+000 1.6915031e+000
Stn 07: 6.9063412e-001 6.5365828e-001 6.4420414e-001 6.4479605e-001
Stn 08: 3.5252224e+000 3.5634836e+000 3.5726562e+000 3.5720643e+000
Stn 12: 1.8683127e-001 1.4931458e-001 1.3020777e-001 1.3079968e-001
Stn 13: 2.5200514e+000 2.5563259e+000 2.5761755e+000 2.5755835e+000
Stn 14: 3.1996903e+000 3.1754926e+000 3.1311856e+000 3.1317775e+000
Stn 16: 7.8569600e-001 7.5505779e-001 7.2803895e-001 7.2744703e-001
Stn 17: 2.3763547e-004 2.4394606e-001 2.8018043e-001 2.8077234e-001
Stn 18: 3.8788962e+000 3.8432842e+000 3.8146641e+000 3.8152560e+000
Stn 19: 5.1009327e-001 4.7987773e-001 4.2097633e-001 4.2689547e-001
```

227

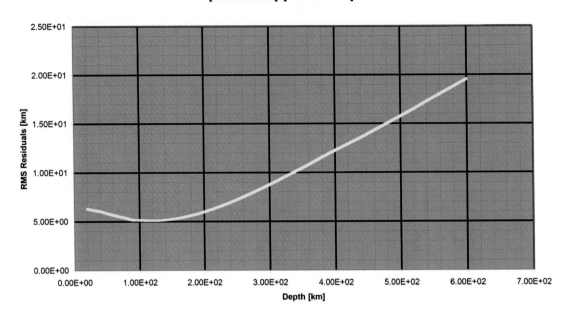

Hypocentre Scan [Lagrange (4)] ak135 (P-wave)
Table Normalisation Lagrange
Interpolation Method: Linear/Linear
H[d] = 118.890 km
[Consensus H[d] = 119.760 km]

Figure H04

```
Station 01: Take Off Angle 1.4374636e+002 degrees.
Station 03: Take Off Angle 1.3688759e+002 degrees.
Station 04: Take Off Angle 1.4372832e+002 degrees.
Station 05: Take Off Angle 1.4035144e+002 degrees.
Station 06: Take Off Angle 8.7648486e+001 degrees.
Station 07: Take Off Angle 1.4091833e+002 degrees.
Station 08: Take Off Angle 1.4250267e+002 degrees.
Station 09: Take Off Angle 1.3956755e+002 degrees.
Station 10: Take Off Angle 1.2762540e+002 degrees.
Station 12: Take Off Angle 1.4155066e+002 degrees.
Station 13: Take Off Angle 1.4012955e+002 degrees.
Station 14: Take Off Angle 1.0708515e+002 degrees.
Station 16: Take Off Angle 9.2201353e+001 degrees.
Station 17: Take Off Angle 1.3697718e+002 degrees.
Station 18: Take Off Angle 1.3912827e+002 degrees.
Station 19: Take Off Angle 8.4456296e+001 degrees.
```

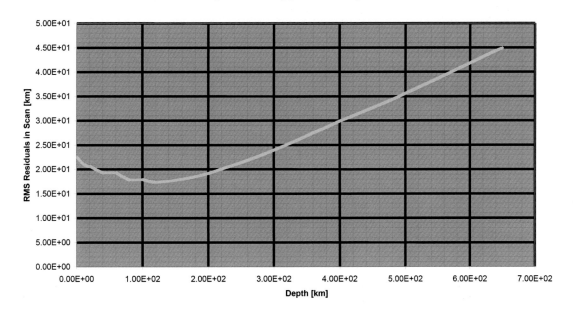

Figure H05

```
Station 00: Take Off Angle 1.2326024e+002 degrees.
Station 01: Take Off Angle 1.4308489e+002 degrees.
Station 02: Take Off Angle 9.6236829e+001 degrees.
Station 03: Take Off Angle 1.3837211e+002 degrees.
Station 04: Take Off Angle 1.4307018e+002 degrees.
Station 05: Take Off Angle 1.4048643e+002 degrees.
Station 06: Take Off Angle 8.7309991e+001 degrees.
Station 07: Take Off Angle 1.4089787e+002 degrees.
Station 08: Take Off Angle 1.4209148e+002 degrees.
Station 09: Take Off Angle 1.3989903e+002 degrees.
Station 10: Take Off Angle 1.2763618e+002 degrees.
Station 11: Take Off Angle 1.2943484e+002 degrees.
Station 12: Take Off Angle 1.4136388e+002 degrees.
Station 13: Take Off Angle 1.4030847e+002 degrees.
Station 14: Take Off Angle 9.4871939e+001 degrees.
Station 15: Take Off Angle 9.4991563e+001 degrees.
Station 16: Take Off Angle 8.9725393e+001 degrees.
Station 17: Take Off Angle 1.3852669e+002 degrees.
Station 18: Take Off Angle 1.3976257e+002 degrees.
Station 19: Take Off Angle 8.4569909e+001 degrees.
```

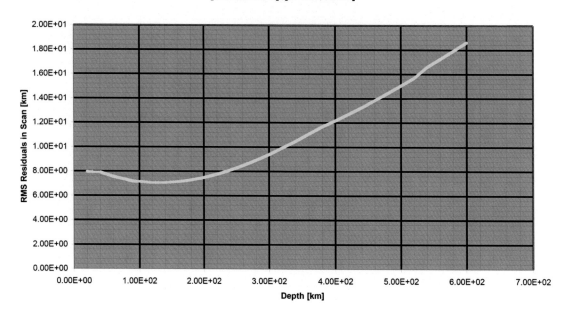

Figure H06

```
Station 00: Take Off Angle 1.2625976e+002 degrees.
Station 01: Take Off Angle 1.4360344e+002 degrees.
Station 02: Take Off Angle 1.1118541e+002 degrees.
Station 03: Take Off Angle 1.3654320e+002 degrees.
Station 04: Take Off Angle 1.4358530e+002 degrees.
Station 05: Take Off Angle 1.4018027e+002 degrees.
Station 06: Take Off Angle 8.8192399e+001 degrees.
Station 07: Take Off Angle 1.4075835e+002 degrees.
Station 08: Take Off Angle 1.4235333e+002 degrees.
Station 09: Take Off Angle 1.3938988e+002 degrees.
Station 10: Take Off Angle 1.2775670e+002 degrees.
Station 11: Take Off Angle 1.2985427e+002 degrees.
Station 12: Take Off Angle 1.4139522e+002 degrees.
Station 13: Take Off Angle 1.3995834e+002 degrees.
Station 14: Take Off Angle 1.0729662e+002 degrees.
Station 15: Take Off Angle 1.0763334e+002 degrees.
Station 16: Take Off Angle 9.2912782e+001 degrees.
Station 17: Take Off Angle 1.3663726e+002 degrees.
Station 18: Take Off Angle 1.3902706e+002 degrees.
Station 19: Take Off Angle 8.4547853e+001 degrees.
```

Hypocentre Scan [Lagrange (4)] ak135 (P-wave)
Table Normalisation Linear
Interpolation Method Linear/Linear
H[d] = 120.05 km
[Consensus H[d] = 120.05 km]

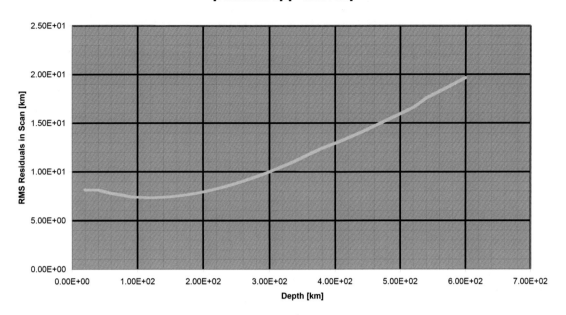

Figure H07

```
Station 00: Take Off Angle 1.2599945e+002 degrees.
Station 01: Take Off Angle 1.4373999e+002 degrees.
Station 02: Take Off Angle 1.1120704e+002 degrees.
Station 03: Take Off Angle 1.3688008e+002 degrees.
Station 05: Take Off Angle 1.4034358e+002 degrees.
Station 06: Take Off Angle 8.7612451e+001 degrees.
Station 07: Take Off Angle 1.4091103e+002 degrees.
Station 10: Take Off Angle 1.2762664e+002 degrees.
Station 11: Take Off Angle 1.2997830e+002 degrees.
Station 12: Take Off Angle 1.4154360e+002 degrees.
Station 13: Take Off Angle 1.4012169e+002 degrees.
Station 14: Take Off Angle 1.0740053e+002 degrees.
Station 15: Take Off Angle 1.0772586e+002 degrees.
Station 16: Take Off Angle 9.2331304e+001 degrees.
Station 17: Take Off Angle 1.3696818e+002 degrees.
Station 18: Take Off Angle 1.3910507e+002 degrees.
Station 19: Take Off Angle 8.4353265e+001 degrees.
```

Annex I

Derivation of the Calculation for the Surface Geodesic Curve Constant

The square of the distance between two neighbouring points is

$$ds^2 = dx^2 + dy^2 + dz^2 = \sum_{u,v} h_{u,v}^2 \, dudv$$

We have

$$dx^2 = \left(\frac{\partial x}{\partial u}\right)^2 (du)^2 + 2\frac{\partial x}{\partial u}\frac{\partial x}{\partial v} dudv + \left(\frac{\partial x}{\partial v}\right)^2 (dv)^2$$

and similarly for dy and dz.

there is an array of scale factors (the metric):

$$h_{u,u}^2 \; h_{u,v}^2$$
$$h_{v,u}^2 \; h_{v,v}^2$$

Equating like with like and assuming orthogonality:

$$h_{u,u}^2 = \left(\frac{\partial x}{\partial u}\right)^2 + \left(\frac{\partial y}{\partial u}\right)^2 + \left(\frac{\partial z}{\partial u}\right)^2 \left(\triangleq h_u^2\right)$$

$$h_{v,v}^2 = \left(\frac{\partial x}{\partial v}\right)^2 + \left(\frac{\partial y}{\partial v}\right)^2 + \left(\frac{\partial z}{\partial v}\right)^2 \left(\triangleq h_v^2\right)$$

This leads to

$$h_u^2 = r_e^2 sin^2(u) + r_p^2 cos^2(u)$$

$$h_v^2 = r_e^2 cos^2(u)$$

232

We also have

$$ds^2 = h_u^2 \cdot du^2 + h_v^2 \cdot dv^2$$

$$\frac{ds^2}{du^2} = \left(\frac{ds}{du}\right)^2 = h_u^2 + h_v^2 \cdot \left(\frac{dv}{du}\right)^2$$

$$\sqrt{\left(\frac{ds}{du}\right)^2} \cdot du = \sqrt{\left(r_e^2 sin^2(u) + r_p^2 cos^2(u) + r_e^2 cos^2(u) \cdot V^2\right)} \cdot du$$

$$= F(u,V) \cdot du; \quad V = \frac{dv}{du}.$$

In short, the integral, $\int_{\lambda_1}^{\lambda_2} F(u,V) \cdot du$ (which gives arc length in Cartesian three-space as a function of displacements in latitude and longitude, (u,v), and whose end points are fixed) is *stationary for weak variations*. This is the case, at least, if v satisfies the condition in the differential equation below.

So by theorem 3, at (16), and setting:

$$\frac{\partial F}{\partial V} = k,$$

we can generate required geodesic paths on integration *if we can find* a v that can satisfy this differential equation. However, a constant k can be found, which will correspond to the requirements of the function V, in the above, as

$$\frac{dF}{dV} = \frac{r_e^2 \cdot cos^2(u) \cdot V}{\sqrt{\left(r_e^2 sin^2(u) + r_p^2 cos^2(u) + r_e^2 cos^2(u) \cdot V^2\right)}} = k$$

$$r_e^4 cos^4(u) \cdot V^2 = k^2 \left(r_e^2 sin^2(u) + r_p^2 cos^2(u) + r_e^2 cos^2 \cdot V^2\right)$$

$$V^2 = \left(\frac{dv}{du}\right)^2 = \frac{k^2 \left(r_e^2 sin^2(u) + r_p^2 cos^2(u)\right)}{r_e^2 cos^2(u) \cdot \left(r_e^2 cos^2(u) - k^2\right)}$$

If k is to remain constant, then this integral equation must hold:

$$v = \frac{k}{r_e} \cdot \int \frac{1}{cos(u)} \sqrt{\frac{r_e^2 sin^2(u) + r_p^2 cos^2(u)}{r_e^2 cos^2(u) - k^2}} \cdot du$$

The constant k can be found by numerical means for a particular geodesic trajectory, defined by its end points as (latitude, longitude) pairs. (For example, this can be achieved by searching for a crossing and then performing a linear interpolation.)

Finally the process

$$\phi_i = \frac{k}{r_e} \int_{\lambda_i}^{\lambda_i} \frac{1}{\cos(u)} \sqrt{\frac{r_e^2 \sin^2(u) + r_p^2 \cos^2(u)}{r_e^2 \cos^2(u) - k^2}} \cdot du + \phi_1,$$

can be evaluated numerically to give the trajectory of the geodesic associated with the particular value of k: $\{(\lambda, \phi)_i\}$. This will occur on the surface of the spheroid, between and including the end points— $(\lambda, \phi)_1$ and $(\lambda, \phi)_2$ as an explicit function of u, the latitude.

Annex J

Grid Scan for Global Minimum Using a "Spider"

The angle of attack here is approached in four stages. The basic idea is to note that, if the vector from the subspace

$$\underline{v} = \begin{bmatrix} \delta t \\ \overline{c}_0 \end{bmatrix}$$

is known, then the set of fixing equations can become

$$\underline{x} \cdot \underline{a}_i = \rho_i\left(\delta t, \overline{c}_0\right)$$

in other words, a set of hyper-planes in three-space. For any known $\underline{v}$ the vector $\underline{x}$ can be evaluated by an explicit least-squares method. Then, if the $\left(\delta t, \overline{c}_0\right)$ pair is the right one, the resulting Cartesian vector should lie on the spherical surface. When this resulting $\underline{x}$ and its generating $\underline{v}$ are placed in the cost function

$$S_n = \sum_{i=0}^{n-1}\left(\underline{x} \cdot \underline{a}_i - R^2 \cos\left(\frac{\overline{c}_0\left(t_i - \delta t\right)}{R}\right)\right)^2$$

a minimum will be formed, if not a zero.

1. The subspace referred to above is reduced to the region within the limits of

$$\delta t \in \left[t_{\min}, Zero\right]$$
$$\overline{c}_0 \in \left[\overline{c}_{0_{\min}}, \overline{c}_{0_{\max}}\right]$$

Physics will tell us what is sensible to use for the limiting velocities. Since we are only considering first-time arrivals, and, by definition, the time to origin, δt, is a negative or zero quantity only:

$$t_{\min} = -\frac{\pi R}{\overline{c}_{0_{\min}}}.$$

There is one caveat. When all the sensors (seismometers) are placed on a great circle, then there is no unique solution to the linear system (1, 2). If the direction cosines, $\underline{\lambda} = (\lambda, \mu, \nu)$, are part of this homogeneous hyper-plane equation containing the sensor locus we can write

$$c_i = -\frac{\mu}{\nu} b_i - \frac{\lambda}{\nu} a_i$$

And the left-hand side (1, 2) loses rank by one. In fact, there are two possible solutions placed symmetrically on the sphere (earth), on either side of this plane. The surfaces formed from the values of

$$S_n = f\left(\{t_i\}, \delta t, \overline{c}_0\right)$$

are clear smooth and instructive, showing for each given sensor grouping and their corresponding events, ridges and valleys in the regions where "true" and "spurious" fixes will be found.

2. Each possible velocity has its own t_{min}. The lower bounds for time become a hyperbolic function of velocity. The limits for the scan are now

$$\delta t \in \left[-\frac{\pi R}{\overline{c}_0}, Zero\right]$$

$$\overline{c}_0 \in \left[\overline{c}_{0_{min}}, \overline{c}_{0_{max}}\right]$$

This assumes that the maximum possible value for a first arrival is

$$|\delta t| = \frac{\pi R}{\overline{c}_0}$$

This cannot be so, since its travel past the lead sensor (time origin) has taken the maximum value in the time set $\{t_i\}$, as well as $|\delta t|$, to complete. Thus,

$$\max\{t_i\} + |\delta t| \le \frac{\pi R}{\overline{c}_0}$$

Hence, the least possible and feasible t_{min} for first arrivals at any velocity level is given by

$$\delta t \in \left[-\frac{\pi R}{\overline{c}_0} + \max\left(\{t_i\}\right), Zero\right]$$

3. These observations speed up the explicit least-squares scan in the two-space occupied by the $\underline{\nu}$. And there is a tendency to isolate the required valley in which the true minimum is to be found. We can set up a gradient descending scan in the form of a "spider". The spider will consist of nine neighbouring grid points, in a three-by-three lattice, for which the cost function, S_n, has

been calculated. The spider will then centre itself onto that grid point within its lattice, which has the smallest value. At each of the relative grid points in the spider, a candidate $\underline{x}_{0_{(i,j)}}$ is calculated and stored separately from its use in calculating its corresponding $S_{n_{(i,j)}}$ at the lattice point (i, j). In general terms, the spider stops when its centre lattice point $(2, 2)$ corresponds to a minimum value of the $S_{n_{(i,j)}}$.

4. We refine one stage further. We note the expression at (0, 7) and above and write

$$-\frac{\pi R}{\overline{c}_0} + \max\left(\{t_i\}\right) \geq Zero.$$

We can limit the scan again if we find a velocity such that

$$\overline{c}_0 = \frac{\pi R}{\max\left(\{t_i\}\right)}; \quad \overline{c}_0 < \overline{c}_{0_{max}}; \quad \overline{c}_0 > \overline{c}_{0_{min}}.$$

Printed in the United States
By Bookmasters